THE COMPLETE ZAHA HADID

扎哈·哈迪德全集

［英］扎哈·哈迪德　［美］亚伦·贝斯基　编著

梁雪　译

江苏凤凰科学技术出版社

THE COMPLETE ZAHA HADID

扎哈·哈迪德全集

图书在版编目（CIP）数据

扎哈·哈迪德全集 /（英）扎哈·哈迪德，（美）亚伦·贝斯基编著；梁雪译. -- 南京：江苏凤凰科学技术出版社，2018.3
 ISBN 978-7-5537-8863-0

Ⅰ.①扎… Ⅱ.①扎…②亚…③梁… Ⅲ.①建筑设计-作品集-英国-现代 Ⅳ.①TU206

中国版本图书馆CIP数据核字(2017)第329176号

First published in the United Kingdom in 1998 by
Thames & Hudson Ltd, 181a High Holborn, London WC1V 7QX

This revised and expanded edition 2017
The Complete Zaha Hadid © 1998, 2009, 2013 and 2017
Thames & Hudson Ltd, London
Text/images copyright © 1998, 2009, 2013 and 2017 Zaha Hadid Architects
Copyright details for illustrations appear in the Picture Credits, p. 319
Introduction copyright © 1998, 2009, 2013 and 2017 Aaron Betsky

扎哈·哈迪德全集

编　　著	[英] 扎哈·哈迪德 [美] 亚伦·贝斯基
译　　者	梁　雪
项目策划	凤凰空间 / 李文恒
责任编辑	刘屹立　赵　研
特约编辑	李文恒
出版发行	江苏凤凰科学技术出版社
出版社地址	南京市湖南路1号A楼，邮编：210009
出版社网址	http://www.pspress.cn
总 经 销	天津凤凰空间文化传媒有限公司
总经销网址	http://www.ifengspace.cn
印　　刷	上海利丰雅高印刷有限公司
开　　本	889 mm×1194 mm　1/16
印　　张	20
字　　数	256 000
版　　次	2018年3月第1版
印　　次	2018年3月第1次印刷
标准书号	ISBN 978-7-5537-8863-0
定　　价	298.00元（精）

图书如有印装质量问题，可随时向销售部调换（电话：022-87893668）。

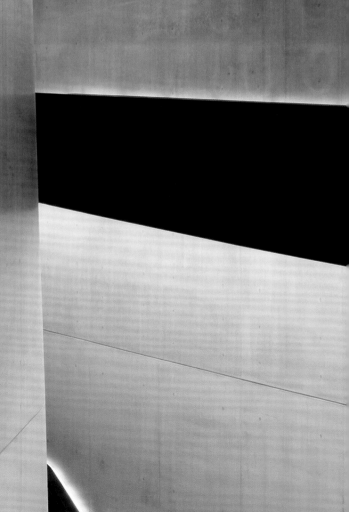

6	简介：在89度之外
17	建筑和项目
259	物品、家具和内部装饰
304	项目信息
320	索引

简介：在89度之外

亚伦·贝斯基（Aaron Betsky）

一方面，电影深化了我们的理解力，使我们更为深刻地洞彻了那些凌驾于我们生活之上的必然法则；另一方面，电影也赋予了我们一片广阔的天地，让我们尽情地去做那些不可思议的事情。都市大街和街边的酒馆、办公室和精心装修的屋子、地铁站和工厂——这一切把我们锁在了深深的绝望之中。终于，电影出现了，它宛若一枚炸弹，以十分之一秒的速度炸开了这个牢笼般的世界；而如今，置身于硝烟沉寂后的废墟之中，我们冷静地踏上了这场历险之旅。特写镜头拓宽了空间，慢动作则延长了幅度。人们放大了抓拍下的场景，虽然放大后的内容变得模糊，我们的眼前所见却变得更为清晰：它以全新的方式向我们展示出所拍摄对象的结构构成。同理，慢动作不仅向我们展现了运动的普遍特性，还揭露了某些深藏其中的未知特性；"慢动作看起来绝非只是简单地将动作的速度放慢，而是展现出一种流畅的、浮动的、灵异的效果"。显然，一种与众不同的特性曾不为人们的肉眼所见，而如今却在镜头前尽情地展现出来——这也许仅仅是因为人们有意地探索出了一个世界，从而代替了我们曾漫无目地穿梭于其中的那个世界。[1]

十分之一秒的爆破

扎哈·哈迪德曾是一位伟大的电影摄影师。她每时每刻都像一架摄影机一般捕捉着周遭的世界。她以慢动作、摇镜头、抓场景的方式捕捉着城市，她眼中的城市带有蒙太奇式的镜头切换，以及讲故事般的节奏。她从我们的现代城市构造中找出隐匿的部分，然后将其作为一个故事编入她的乌托邦。她的探索十分大胆，她会时而放慢时而加快日常生活的节奏；而作为一种表现方式，她会用环境来诠释建筑的外形。她塑造了十分之一秒的爆破。

但这并不意味着她不是一位建筑师。她一直都希望能够建造些什么，而她所创造的意向——她早期所绘制的作品，以及她笔下那些呼之欲出的构造、那些天生适用于镜头的形状——连同其他因素一起将她推入了建筑业。然而，她并不想仅仅将一个毫不相关的物体置于某处空荡荡的场所里。相反，她所创造的建筑物本身有着极强的凝聚力，它们散发着延伸的力量。从建筑物的项目规划到先进的基础设施，她将所有的能量都凝滞于建筑物之中，从而赋予它充分的存在价值。因其具备高度凝聚的特点，她的建筑物能够自由地突破外在的限制，从而创造无拘无束的空间。这里也许曾经是（或者说将成为）高墙林立、管道遍布的私人活动场所，而如今碎片状和平面的结构却将一切禁锢打破，向我们展现出一个难以置信的空间。

哈迪德以类似的方式塑造着她的建筑生涯。她将少女时期织毛毯的记忆融入了在伦敦建筑联盟所接受的教育之中。她以20世纪早期艺术家的方式塑造空心砌块，以此盖起了她那一座座抽象记忆的殿堂。她画下了城市的活力，而周遭山水那浓重的轮廓则如斗篷一样将她层层包围。于是她借着山水的力量，以山水的轮廓为起点探索着未知的领域，而她笔下那些棱角分明的形状正表现着这种未知。

人们也许会说，扎哈·哈迪德是一位现代主义者。作为对新事物的标榜，她所设计的阁楼具备技术精湛的内核。[2] 哈迪德对建筑类型学、实用秩序、隐含假设以及地心引力不感兴趣：她相信我们能够并且应该构造一个更美好的世界，在这个世界中，最推崇的便是至高无上的自由。我们可以将自己从过去中解放出来，我们可以摆脱社会习俗、物理法则的束缚，让自己的身体获得自由。对于哈迪德这样的现代主义者来说，建筑一直都是这个世界中碎片化的构成。

现代主义的三种方式

一贯而言，此类现代主义风格表现为三种方式。首先，它的坚守者信仰新的结构。在技术的协助下，一位优秀的现代主义者设想我们可以更为高效地利用我们的资源（包括我们自身在内），去最大限度地创造无论是空间方面还是价值方面的盈余。这种"过剩"恰恰表现了永恒、未来和乌托邦的真实，它有着英雄主义的色彩。它无影无形，当它经过我们周遭的时候，早已将自身的形象归于虚无。其次，现代主义者相信新的视觉方式。也许世界已经是日新月异的，但我们却没有认识到它的活灵活现。我们仅仅是看到了教育所要求我们看到的东西。如果我们愿意以崭新的方式去看，单是凭借这一举动，我们就可以改变整个世界。我们需要睁开我们的眼睛、竖起我们的耳朵、打开我们的心灵去感受我们存在的实质。而当我们这样做时，我们已然获得了自由。第三，现代主义者期望展现现代性的实质。通过将以上所讲述的两个方式融合在一起，他/她能够将我们所捕捉到新的感觉转化为我们所创造的形状。这些形状是现实的原型，事物在其中被重新排列和溶解，直到除了新事物之外的一切都消失不见。通过以新的方式展示新的事物，我们能够用自己的眼睛构造一个新的世界，从此栖息在那里。

正是这第三种方式塑造了扎哈·哈迪德的作品。她并没有创造出新的建筑形状或技术；她以极端的表现方式向我们展现着世界那崭新的一面。在主体与客体的消解中，她找到了现代主义的根本，她将其搬上了现代风景的舞台，并将其重新塑造为我们可以大胆漫游的空间。

这种现代主义的典范至少可以追溯至巴洛克时期。那时，主体和客体第一次失去

匹克项目

了它们那不容置疑的权威。人类的肉体不再置身于充满罪恶的世界中、不再立于上帝的面前；唯有现实在不断地延续，而自我则交融于其中：

> 物质因而向我们展现出多孔的、弹性的，或是洞穴般的质地。物质之中无空洞可言，洞穴永无止境地包含于其他洞穴之中，无论多么娇小的个体中都包含着一个世界，这个世界中贯穿着不规则的通道，液体在这个世界之中流动，带着不断升腾的雾气在这个世界里穿梭，而宇宙的总体则如同"一池物质一般，充满了形态各异的涌流和浪花。"³

建筑物试图以多样的形状表现这种流动的能量：

> 巴洛克风格创造了无数的作品和方法。问题不在于如何完成一次交叠，而在于如何使其延续，使其表现在天花板上，使其无限地延伸下去……交叠影响具体的形状，并使其呈现在人们面前。交叠创造出表达的方式和设计的样式，它创造出作为起源的元素以及蔓延无尽的曲线，它赋予弧线以独特的游走姿态。⁴

显然，工业革命塑造了一个如此混乱的世界，它将存在的意义和价值从每件物品、每个人身上移除，然后将其卷入资本的洪流之中。因此，玻璃、钢筋和混凝土让建筑物变得千篇一律，建筑物如洪流一般环绕着最后几幢形态独特的遗迹，将它们掩埋于不断积累的消费品之中。正是扎哈·哈迪德创造了这些涌流。

行外人的融入

然而，哈迪德的作品却不止于具备现代性的本质。生于伊拉克，她表示自己在少女时期曾痴迷于波斯的地毯。地毯上的花纹错综复杂，微妙到超乎人们的想象，诉说着两只手如何相互配合着将现实性诉诸于物体表面、将简单直接的空间变换为繁茂苍翠的世界，从而使美感能够为感观所捕捉。值得注意的是，这样的作品也往往出自女人之手。⁵

在倾听哈迪德作品中那不断展现的叙事时，人们也可以将其与中国以及日本的卷轴画相比较。当日常生活中的一举一动积累起来，便不断地改变了我们的现实性。现代主义提议我们从积累的日常行为中寻找感觉，而不是将某种特定的秩序注入物体之中。这种工作原理广为卷轴画画家所知。他们在自己的作品中蜿蜒进出，他们聚焦于微小的细节，他们多次以不同的角度展现场景，他们将独立的元素融入山水风景。他们挥动画笔，使线条彼此呼应，在线条的交织下，美景跃然出现在观者眼前，原本的世界就这样被更改，归还给我们的是一个变换过的世界。

在20世纪早期，艺术家们曾尽情地运用这些传统手法，而他们的艺术创作为哈迪德提供了灵感，使她创造出美轮美奂的建筑砌块。无论是在立体主义、表现主义还是至上主义之中，抽象的碎片往往被置入叙事结构之中。这些艺术家们轰炸了他们所在的那个世界——杜尚的《下楼梯的裸女》激发了扎哈·哈迪德最初的灵感。

伦敦的建筑联盟曾给了哈迪德最直接的启蒙，当她在那里学习时，这所学校正处于巅峰时期。当时，这里是世界建筑实验的中心。这所学校建于阿基格拉姆学派的遗迹之上，这里的学生和教师曾包括彼得·库克（Peter Cook）、雷姆·库哈斯（Rem Koolhaas）、伯纳德·屈米（Bernard Tschumi）以及奈杰尔·科茨（Nigel Coates），他们将现代世界中的动乱转换为自己作品中的主题和形式。通过讲述我们周遭变幻的日常，他们敢于再一次以现代主义的方式，试图去捕捉蕴含于其中的能量。他们试图赋予现代性以形状，而当他们这样做的时候，他们也为自己的尝试开启了一个叙事的视角。无论作品专注于描绘奇闻逸事、表现晦涩复杂（屈米），还是用于展现神秘的风格杂糅（库哈斯），抑或是一份个性宣言（库克），他们都赋予了意向以多样的视角和气势恢宏并富于表现的形式，以及技术精湛的框架，而这些意向所表现的事物则具备描述的性质而非界定的目的。

浓缩的杂糅

正是在这样的境遇中，扎哈·哈迪德的作品成形了。她所设计的第一项引人注目的工程是一架建造在泰晤士河流之上的桥梁［马列维奇的建构（Malevichi's Tektonik, 1976—1977, 第18页）］。作为她的毕业设计，这项工程无疑受惠于她与雷姆·库哈斯的来往——她曾与大都会建筑事务所合作长达3年之久——在库哈斯的影响下，她学会在作品中突出一个去除冗余、唯留精华的几何结构，这种方式直白地向马列维奇的至上主义作品致敬。她所绘制的桥梁看起来很像马列维奇的那一架飞机，看起来既可以是一群雕塑，也可以是一排房子。她故意使笔下的意向具备中性的特质，用当时在建筑事务所非常流行的一个词来说，她将建筑物看作是"社会凝聚器（social condenser）"。这座建筑物本身是一幢现代主义的阁楼，它向下将自身折叠起来，以使得不同的组合元素（实际上她并没有刻意地表现这些元素）与彼此亲密接触。然而，看到这项工程，令我们这些观者震惊的既不是它在实用层面的追求，也不是它对于以往杰作的引用，而是意向本身：它以一种不屈不挠的新颖姿态跃然于纸上，占据

特拉法加广场大厦

了我们的全部视线。

在毕业之后的数个项目之中，哈迪德继续发展自己的叙述风格，并将其进化为一种更为完整的空间语言。她为自己的兄弟所设计的一座公寓——59号伊顿普雷斯公寓（59 Eaton Place, 1981—1982, 第21页）能够直接唤起一些爱尔兰共和军轰炸的记忆，当时炸弹就落在这附近。这幅图所描述的本身就是一场爆炸，置于图片之上的元素本身就是一些碎片，它们被定格在以最现代的方式释放能量的那一瞬间。它将成为哈迪德所创造建筑中的一个重要主题，物体在其中被压缩，而城市的形状则被缩影至家具之中。这些画中的碎片作为流行艺术元素，重新回到它们原本所在的位置上；而一座舞台早已被搭起，等待着上演一座现代城市的重建。哈迪德将59号伊顿普雷斯公寓的草图进一步拓展成了美轮美奂的哈尔金广场（Halkin Place, 1985, 第28页）。当观者站在建筑物高处俯瞰时，会发现自己直入云霄，早已将城市中那一排排低矮房屋的屋檐甩在身下，从而以彼得·潘（Peter Pan）式的视角观看着这座城市如何在天地接合处遁向两边；在这样的观看视角之下，哈迪德作品中那些来自于现代主义乌托邦中的碎片重获了它们曾有的历史形态。

通过为爱尔兰总理所做的新住宅提案（1979—1980, 第20页），哈迪德将拼贴艺术引入了自己的作品之中。表现性的元素（瓦片、球面、砖块）占据了一个简洁的立方体，而一面长长的墙蜿蜒而过，打开了这组建筑的故事。哈迪德并未向我们讲述这项规划或者是这一地点，而是唤醒了这一处所所特有的世界主义性质；她并未为我们编制情节，而是直接为我们设立了场景。而说到位于伦敦的特拉法加广场大厦（GrandBuilding, Trafalgar Square, 1984, 第25页至第27页），她为这个项目所做的提议则综合了她许多作品的特色，向我们展现出她对城市景观的想象力。这幅绘图是一幅双联画，它至少从5个视角描述了这座建筑物。它同样表现出城市如何以一种从正面朝上和上下颠倒的方式脱离于其自身的束缚，从而创造出一种不稳定的效果，让人分不清究竟哪里是倒影，哪里是绘画中着重表现的实体。通过将埃舍尔（Escher）绘画中的巧妙与构成主义者作品中的抱负相结合，哈迪德一层一层地分解了城市。

对于这种表现方式所遵循的原理，哈迪德给出了提纲挈领的解释：在特拉法加广场大厦的设计中，其活动和形状被融入质地紧密的构图之中，从而释放一层又一层的开阔空间，以使得城市和这座建筑相互融合，而建筑物那咄咄逼人的形状则向外伸入城市。她从接合处打开了这座城市，而我们所感受到的

现实便与这新设计（新建筑物）在接合处相遇，这种表现手法便是她的绘画中所不断重复的一个主题。在此处所提及的设计中，她便将这一艺术手法植入了意向自身的内部，从而将特拉法加广场继续置于它那被过多游客所充斥的现实之中，而她所设计的建筑物则伫立于一个乌托邦里——那里充满了想象，这想象可以往任一方向延展。

这些早期作品总体上以两种形式呈现。第一种形式集中体现在一张绘图中——世界（89度）[The Word（89 Degrees），1983，第24页]。在这幅画中，哈迪德将她的艺术设计放置在我们所在的这颗星球上的现实世界里，而我们则可以从一架飞机上或者是一颗射向太空的导弹上欣赏这些艺术品。随着地球的旋转，其上的风景映入空中，形成新的几何图形的碎片。现实世界变成了哈迪德的天地，在那里地球引力消失了，视线被扭曲了，线条互相靠拢交织，那里没有尘世间对于尺度与行为的定义。这幅画所展现的并非具备实用性质和形式的特定风景，各种可能出现的构造群集在一起，它们共同塑造出一片名副其实的风景：人类用双手塑造出了这一片天地，以艺术的方式描绘着我们生存于其中的自然环境。

第二种总体的形式让哈迪德声名鹊起。她赢得了参加中国香港太平山顶竞赛的资格（1982—1983，第22页至第23页）。她向数以万计的建筑师和设计专业的学生（包括本书的作者之一）证明了她所开拓的这些技巧是一种新的建筑形式。这项设计本身便是它所在场所的缩影，它本身便凝聚着其他所有设计的特色：它们丢弃了世俗意义上对于存在的要求，它们宁愿让形状以纯粹随心所欲的方式聚集在一起。作为一种设施，这座建筑致力于以社会似乎容许的方式取悦并规训自身。

哈迪德在建筑中以管道表现着她的设计和场地。管道互相叠放于彼此之上，如圆木一般在施工场地堆积在一起。它们在工地上垂直堆积，以悬臂梁的形式构造出层次分明的空间。形状间的缝隙体现着山峰作为一个社交俱乐部的功能：各种活动在这里交替进行，人们在这里相互结识；而横梁则移动着，似乎意在捕捉和固化身体移动的轨迹。这座建筑让人类和山脉相互审视。它不仅仅"为山峰加冕"，还将这座山峰一把折断，让当代的我们能够像远古的巨人一般，与山峰交战。

哈迪德将这一景象铺设在一张非常大的画布上，画作似乎在企及山峰本身的宏大规模。虽然作为一位建筑师，她看重强调自己建筑物中的原理性，但图画本身却让零件和碎片四处飞扬，它轰炸了自身所在的场所和自身所代表的设计。在一幅画作中，哈迪德将俱乐部的元素融入香港闹市之中，而在中心的摩天大楼之下，抽象的飞机在盘旋飞行，它们实际上已经融入了山峰的建筑构造。在这些作品中，哈迪德展现着一座座建筑物，它们将抽象的几何图形聚集在一起，以艺术的手法表现着那座城市以及任何一座城市中的风景。这些新颖的碎片大厦指向了一种更为开阔、剧烈并且多变的空间安排。

在设计的海洋上前行

在接下来的10年里，哈迪德在世界各地拓展着她的主题，将它们表现在她的建筑、设计以及提案中。其中，许多建筑、设计和提议都在德国完成。其中包括柏林国际建筑展的2号住宅（IBA-Block 2，1986—1993，第33页）以及位于莱茵河畔魏尔的维特拉消防站（Vitra Fire Station，1990—1994，第52页至第55页）。前者确立了用于设计特拉法加广场大厦的基本形状，而后者则标志着她的艺术生涯的新阶段。

柏林的维多利亚城市区域（Victoria City Areal，1988，第40页至第41页）、汉堡的港湾道住宅小区（Hafenstraβe Development，1989，第44页至第45页）以及位于杜塞尔多夫莱茵河港口的佐霍夫3媒体公园（Zollhof 3 Media Park，1989—1993，第50页至第51页）具备一些相似的特点。而如今，这些特点已经成为了哈迪德的代表性形状，比如，围绕着一个古怪空洞而建的一圈阁楼、伸入到建筑物内部的公共空间以及向城市内部伸去的形状。在过去的那些年里，这些形状几乎具备了与众不同的特色，然而它们本身也在不断地变换着特质。它们变得更轻、更透明、更具层次感。在某种程度上，这种变化是设计变得更宏大的结果。在大多数情况下，这种变化也是设计变得更普遍的效果。那些单一的办公楼和公寓大楼不具备杂糅的元素，所以赋予它们以描述性的性质。

人们也可以感觉到，哈迪德的关注点已发生了变化。早期的时候，哈迪德将互不相关的元素组合在一起形成拼贴艺术，而如今她的形状似乎以一种独立的动态而存在。对于哈迪德来说，这是她将自己的作品看作风景的结果，换言之，她在塑造大地。在设计特拉法加广场大厦的时候，她首次提出了凝滞和伸展的手法。在设计维多利亚城市区域的时候，她依旧遵循着这两种手法；而在设计杜塞尔多夫莱茵

维特拉消防站

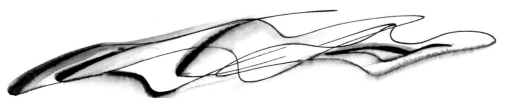

伊斯兰艺术博物馆

河港口的大型复杂设施时，它们看起来却像现代主义冰山的碎片，而裂缝的边缘则看起来是在预示着某种未完结。这些裂缝展现着每一幢建筑的不完整性。在杜塞尔多夫项目的设计中，这座复杂的综合设施具备各种各样的功能区，而这些功能区则具备着相似的形状。桥梁、通道以及公共建筑与这些被戛然截断的功能区一一相连，它们在空间中自由地伸展着结合在一起。无论是在公共区域还是在办公大厦里，每一处都隶属于同一片布满独特形状的宇宙之中。

哈迪德使用色彩的方式也开始发生变化。大都市（Metropolis, 1988，第38页）有着热情的意向，由颜色所编码的碎片仍然占据着两座柏林大厦，但其他德国设计的着色则明显柔和了许多。这也许是因为如今玻璃主宰着建筑材料，又也许是因为德国的城市环境变得相对暗淡了。但这似乎也标志着哈迪德的调色板已经冷静了下来：色度和色调、延伸的形状层次以及柔化的强度代替了以碎片为素材的拼贴艺术。

这些变化在维特拉消防站的设计中得到了最大的发挥。当你在弗兰克·格里（Frank Gehry）那全白色的著名博物馆里看见这项设计的时候，你一定会一眼认出这座建筑物那船头一样的形状。在现实中——哈迪德的草图也表明了这一点——从抽象层面来说，哈迪德将消防站从与它相邻的工厂大厦那里戛然截开，一条蜿蜒的通道从中穿过，它蜿蜒至博物馆并绕着这座综合设施铺设开来。迸发于场景之外的力量冻结了氛围，静谧势不可挡地包围了工厂的外墙。建筑物迎接着沿消防站轮廓而来的风景，而不是以一种敌对的态度伫立在那里。地理环境延伸至建筑物内部，而在建筑物内部，消防车所享用的更大空间则一直延伸到了沐浴区以及休息区，台阶也随着地理环境延伸至第二层。

哈迪德用维特拉项目证明了她可以塑造出一处风景（之后她还曾用位于斯特拉斯堡之外的汽车站证明了这一点）。虽然形状看起来似曾相识，但相比于她早期作品中所创造的集合艺术，她已经又向前发展了很大的空间。从前，她喜欢在大地上搭建建筑物，她喜欢打开新的空间并将形状置于其中，她喜欢让建筑物以攻击性的动态挑战周围的环境。而如今，她借助场地形成她的形状，赋予它们各种各样的功能，并以空间逻辑构造出不朽的建筑。她的建筑物具备了记忆的功能，它们记忆着旷野如何绵延为小山、山洞如何在小山上张望、河流如何流淌过起伏的风景、山峰如何指引着方向。也许哈迪德已经意识到，"十分之一秒的爆破"无法彻底地揭示人类心灵的构造，而作为人类栖息的沉淀物，已被塑造出的环境却能够在本质上揭示人类心灵的构造，它们遵循着与无机界相类似的原则[6]。她发现自由空间并不存在于乌托邦的碎片里，而是存在于对已有事物的探索中。

螺旋的掌控

在设计了维拉特项目之后，螺旋式结构开始出现在哈迪德的作品中。它出现在折叠金属板造型中，这些金属板环绕着1995年国际建筑博览会的蓝图展览馆（Blueprint Pavilion for Interbuild 95, 1995，第78页）；它出现在位于"城市珍宝"的卷曲构件中，而这项设计位于卡迪夫湾歌剧院（Cardiff Bay Opera House）之内（1994—1996，第70页至第73页）；它还出现在一连串紧凑的空间中，该项设计位于维多利亚与阿尔伯特博物馆的锅炉房扩建（Victoria & Albert Museum's Boilerhouse Extension, 1996，第82页）项目中。哈迪德的建筑物在风景中蔓延，随后这些建筑物似乎想将风景据为己有。它们遵循着设计本身的原则将风景囊括起来，然后又借着周围的环境遮蔽或者是圈扩空间。在锅炉房扩建项目设计中，正如哈尔金广场的走廊一样，该项设计中的走廊部分一直延伸到了屋顶之外。在卡迪夫湾歌剧院中，螺旋式结构包围了主厅中的广阔空间；而在蓝图展览馆中，它们将美丽的看台打造成了一个小型建筑物。

虽然哈迪德后期的大部分作品都是大型建筑物，但是她却继续将它们画成通透的建筑体。她不再以金属板构造出沉重的效果，而是建议通过操纵几何图形和结构从建筑物的内部释放出空间。她将注意力投射到无尽的风景之上，将风景看作敞开的延伸体，将风景看作是建筑物内在的一部分。在哈迪德为这些设计所画下的许多草图中，白色的线条遍布于建筑体黑色的表面之上，似乎这些草图预示着无限的可能性，正等待着人们的诠释。她早期设计中的确定性已经让位于对抽象开放性的动态式探索。

在哈迪德为伦敦的哈克尼帝国剧院（Hackney Empire theatre）所做的设计提议中，以及在她为位于辛辛那提城市中的洛伊斯与理查德·罗森塔尔当代艺术中心（Lois & Richard Rosenthal Center for Contemporary Art）所做的设计提议中，这种透明的、如宝石一般瑰丽的手法达到了极致（1997—2003，第97页至第99页）。在这两项设计中，建筑物的外形淡去了，唯留下建筑物的内在与城市活力的交融。这些力量在扶梯与螺旋式建筑物中以更小范围的方式体现。相互折叠与环环相扣的手法、热烈的形状（墙、地板与天花板）与冷静的环境（居住的空间），这一切都如鳗鱼一般以从未有过的稠密密度围绕着彼此，却构成了更为透明且宽敞的结构。

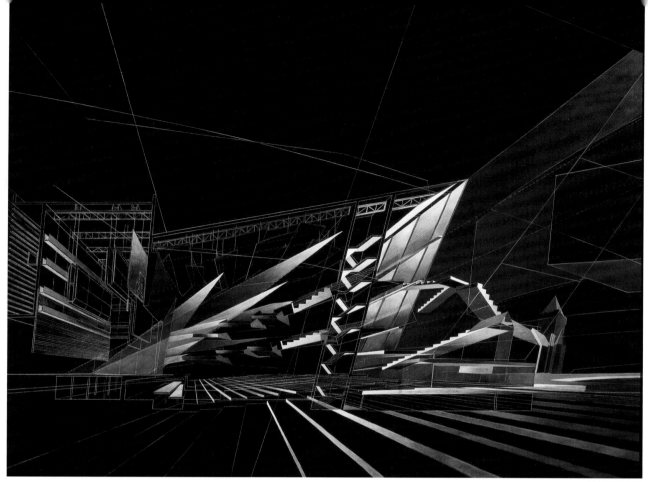

卡迪夫湾歌剧院

　　与此同时，她早期作品中的管状造型也成为了设计中的主要特色。它们被捆在一起构成了位于维也纳的斯皮特劳高架桥（Spittelau Viaducts，1994—2005，第74页至第77页）以及位于伦敦的宜居桥（Habitable Bridge）设计（1996，第84页至第85页）。从某种程度上来说，虽然这些柱状体让人想起了中国香港的匹克项目（The Peak）设计中的平板，但这些管道罗列得更为稠密紧致得多，实际上，已经难以将环绕结构与使用空间分开。它们同样还强调了空间中水平方向的动态，它们与垂直的建筑形状互为补充。在1999年的位于德国园艺展览馆中的设计作品（1996—1999，第88页至第91页）中，管道造型融入了哈迪德先前对于打造风景的兴趣之中，从而创造出了一幅伟大的弯曲的平面图。

走向新风景

　　有一段时期，风景成了哈迪德作品中的主导因素。如果说她所设计的建筑主体越来越具有流动性，那么它们向外延伸的部分也是如此。这样的特点体现在她的许多设计中，诸如位于卡塔尔的伊斯兰艺术博物馆（Museum of Islamic Arts，1997，第92页至第93页），这座建筑物看起来或多或少地像是一圈向外波动的涟漪，它向着上方包围着它的空间蔓延，随后又顺势落到了地面上。庭院中的水渠如一片波斯地毯一般，将空间和实体交织在一起，又如河流或者湖泊一般，在这片土地上来回流淌。作为设计中的固有模式，这些建筑物只能达到符合实际使用需要的高度，但正如她所喜欢的衣服上的花纹一样，或者说正如她所艳羡的潘顿餐椅上的花纹一样，这些建筑散发着其自身所具备的美感，而这美感就蕴含在建筑体本身的流势之中。

　　然而，在这个新世界中，却存在着一种截然不同的现实。关于这种现实，哈迪德在她的一些作品中进行过充分的探索，比如说展于伦敦千禧圆顶馆的思想地带设计（The Mind Zone，1998—2000，第100页）。空间与形状之间错综复杂的交织被风景一般的外表所抚平。在一面墙的作用下，风景的轮廓便被打造成了外悬的船首。

哈迪德在早期的作品中将厚板、船首以及大厦换成了螺旋状以及管状物体。动态以及姿势取代了形状成为了主导因素，而作品也变得更加细腻和抒情。通过打开城市风景、释放现代都市的活力以及创造一个幻影的世界，哈迪德探索了这样一幢建筑物的空间可能性，它的形状具备象征的意义，而它的结构——甚至有人认为——具备文体学性质。

她在这个设计中所展现的方式开始符合她的意图。在过去12年里，哈迪德在绘画以及素描方面所投入的时间已经越来越少了。她的办公室不断扩大，她的佣金不断上涨，她更喜欢投入到工作之中，就像一位文艺复兴时期的大师一样，就像一间工作室的头领一样。她画着草图，画下"所有精确的线条"，而这些线条全部指向她的设计目标;[7] 她的合作者操心着更大规模方面的工作，并负责填满她设计的动态之间的空间。她作品中的细节越来越少，差异与色彩也越来越少。她已从多色彩和重颜料的拼贴艺术走向了单色的涂抹模式，她所绘制的画作上只有白色的线条；这些线条被画在黑色的纸上，宛若未来城市中的幽灵。

银幕珍宝

在伙伴帕特里克·舒马赫（Patrik Schumacher）的督促下，哈迪德开始运用计算机设计与通信技术设计作品，这使得哈迪森能够将自己的开放式画风运用到国际领域之中。先进的软件科技被直接应用到了建筑设计之中，这使得她能够把握已有的风景，将风景按比例展开、旋转、投掷、转向、分割、放慢以及加速，并且赋予了风景以层次分明的复杂性。以前，她只能在自己的草图中表明这一切。基于运算法则和参数计算（舒马赫最终发展了这些理论），这些方式同样也减少了她的作品中的个人探索色彩，而向人们说明：她所想象出来的东西实际上是对内在复杂性以及现代性悖论的真实回应。这同样也意味着，她那间狭小的工作室已经发展成了最新型的生产场所，在这里能够生产出许多遍布全球的大型建筑物。她本人的职能也变成了主导设想家、批评家以及推销员。就像她曾在伦敦基地里所做的那样，她如今在飞机上或者是在工地上指挥着工作室的运转。

国立二十一世纪艺术博物馆

她并不是唯一一个认识到计算机潜力的人。在许多方面，计算机完成了瓦尔特·本杰明（Walter Benjamin）的承诺，尤其是使感知、呈现和认识之间的界限消失。计算机将原本抽象的力量呈现在人们眼前，无论我们做出怎样的选择，它都容许我们塑造并更改现实。随之，它还赋予了这些批评性呈现以可塑的性质。于是，人们操纵着已存在的事物的呈现方式，一种新的事物由此而出现了。

在1983年，当哈迪德将她的世界与我们的世界融合在一起时，她相信自己绘图中的力量能够将我们现实中分崩离析的碎片整合为一种新的事物。在千禧年交替之际，哈迪德已经超越了借助风景的艺术手法而踏入了一个新的领域。这个领域既稠密又开阔、既确定又动荡、既真实又虚幻，它处于89度之外、处于恰当的角度和曲解的几何结构之外、亦处于人类活动被归于形式的事件视界之外。

塑造流动的形状

2004年，哈迪德荣获普利兹克建筑奖。许多人认为，该奖项代表了该行业最高的荣誉。她的面孔出现在流行杂志上,出现在更专业的建筑出版物上。为数百万人所熟识。这样的知名度并非只是徒增热闹。这意味着她可以推销自己乃至自己的作品,她的签名可以提高一项建筑设计的知名度,提高委托人的声誉,使他们更有能力售卖或出租她所设计的公寓或办公空间。在这个过程中,她逐渐形成了自己的标志。

在她所设计的某些作品中,确实存在着专属于她的标志。在生命的最后10年里,她所设计的某些作品上就带有"Z"字形痕迹。它看起来像一条顺势而行的蛇,正蜿蜒地穿过某些业已存在的形状。这个标志以一种更为紧凑包拢的方式,在有限的环境中定义着空间。在意大利罗马的国立二十一世纪艺术博物馆(MAXXI: National Museum of XXI Century Arts, 1998—2009,第106页至第109页)中,以及在德国莱比锡城的宝马工程项目(BMW Plant, 2001—2005,第132页至第135页)中,都能看到蛇的形状在有限的空间中蜿蜒穿梭。在罗马,哈迪德翻新了一些古建筑,将它们作为复杂建筑群的一部分,融入这座被着门口进入内庭、穿过略显拥堵的古老结构、进入仍将被扩建的后部区域。在这个稳固的造型下,阳光从屋顶的构造之间投射下来,而一层层的画廊则向着阳光所在的上方延伸。空间在一层层的画廊之间延伸,拐角处、上下楼梯以及扶手电梯之间、复杂的空间设置上,都给人一种循环往复的美感。这一切都是一种简单的流势所塑造的效果,这一切都创造出了无尽的空间延伸感,让人感觉走过拐角之后又是一番天地。屋顶将这一切聚拢在一起,并且屋顶也成就和守护着这个全新的空间。延展的结构塑造了延展的空间,但这二者并不彼此呼应,是二者之间的对比赋予了建筑以灵性。

在莱比锡城中,这样"Z"字形的造型甚至更明显。在那里,哈迪德不得不围绕着巨大的汽车制造车间工作,而这些车间是别人早已设计出来的。她的任务是塑造出公共空间、办公空间、公共展示区(用于向人们展示汽车是如何制造出来的)以及员工休息区,她将这些看作自己的工作。这里的轨道同样遮蔽着入口,并赋予了它特定的含义,轨道蜿蜒着穿过公共接待区、顺着上升的台阶延伸经过员工的小隔间(台阶之下藏有一间自助餐厅),一直延伸至建筑物的后部;在那里藏有一间体育馆和更多零散的设施,它们为建筑的整体流势增添了最后一抹幽静的色彩。这座建筑中的屋顶架也塑造出了延续的气势,它将所有的元素聚拢在一起。在这座建筑中,高强度的混凝土梁跨越很大的距离,它们在彼此之间塑造出了开敞的空间。与宏大的整体流势相对立的是微不足道的传送带。在传送带上,汽车被从一个生产车间运往另一个生产车间,它们已经被部分的组装好,并且被喷上了油漆。虽然它们看似不值一提,但实际上却赋予了彰显布局性的意义。否则,建筑的整体流势则有可能仅仅停留在表面而不具备深意。

这些建筑宏伟壮阔——位于奥地利因斯布鲁克的伯吉塞尔滑雪台(Bergisel Ski Jump)就是一个具体的实例(1999—2000,第114页至第115页)——它们整体地被塑造成了某种几何图形,这种手法常见于古代建筑,

伊莱和伊迪萨艺术博物馆

而现今仅有最后几座建筑展现出这样的风格。对角图形便出现在位于英国格拉斯哥的河畔博物馆(Riverside Museum, 2004—2011,第170页至第171页)。作为界面元素,它分割并粉碎了整体的规模,以避免建筑呈现出蠢笨而封闭的风格,比如说,位于法国蒙彼利埃的皮埃尔·韦弗斯政府大楼(Pierres Vives, 2002—2012,第146页至第148页)与位于美国密歇根州立大学的伊莱和伊迪萨艺术博物馆(Eli & Edythe Broad Art Museum, 2007—2012,第226页至第227页)。

在她后期所设计的作品中,整座建筑都被塑造成了堆积的球形堆。建筑不再像位于中国的广州大剧院(Guangzhou Opera House, 2003—2010,第156页至第161页)那样,建筑的空间不再被塑造成了盘绕的蛇形,而是被塑造成了一座人工山,山上的软石被铺在了入口处发挥着实际的作用。歌剧院以及其他相关设计利用空间框架以及镀层工艺,将复杂的空间包含在内,这样似乎打破了空间内部的限制。这似乎也向人们展现出,哈迪德的公司越来越依赖计算机来创造出大规模的综合建筑。这些建筑最终实现了结构与空间的相互交织,而这种造型是哈迪德自学生时代便梦寐以求的。

广州大剧院是第一座彻底实现堆积球造型的建筑(这里指的是堆积球的专业建筑意义),它以多个凸出的多面体覆盖了公共项目。主建筑是一座被金属所覆盖的洞穴,在其上和其下,都密布着狭长的空间。它所产生的效果是传统歌剧院的缩影,可以追溯至1875年的查尔斯·加尼尔巴黎歌剧院(Charles Garnier's Paris Opera House)。在这座建筑中,哈迪德的成功之处在于利用数码机器的技术以及空间流动性,创造出不断展现、淡入、淡出、覆盖、环绕观者视线的空间。从外部看,歌剧院给人谜一般的感觉,它展现出无法定义的性质,但又像一只重要的船锚一样,诏示着纪念碑一般的文化意义。但是劣质的建筑材料略显不足。

哈迪德越来地依赖于计算机,她利用计算机展现着延续开来的空间。空间在如波浪一般起伏的建筑物内铺展,而有时这些建筑则被塑造成小球堆积而成的塔。在某些建筑中,

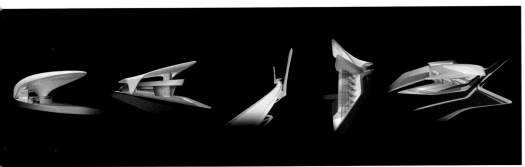

罗德帕克缆车道

蛇状造型重新出现了，它们彼此相互缠绕层叠，比如，阿布扎比表演艺术中心（Abu Dhabi Performing Arts Centre, 2008年至今，第232页至第233页）。蛇状造型均在蛇头部位变换为不同的结构、变换为循环往复的艺术元素。它们围绕着球状的中心空间，塑造出交织的斜坡、横梁以及房间。这样的设计不仅可以使建筑物以物质实体的形式存在，而且可以使人们探索其内部结构的复杂性。这些建筑展现出不可思议的流动性，哈迪德借它们证明了她如何将现实化作不断展现、不断交织的幻境。实际上，这些设计是在最先进的计算机造型设备的帮助下完成的。除此之外，也许有人会补充说，许多这样的建筑也出现在白板景观之中。它们在其中所展现的全部是新的建筑和设施，而不是以建筑的存在来展现隐匿的现实性。

除了蛇形的屋顶和弯曲的形状，哈迪德同样也塑造了行云流水、轻如空气的构造。它们与地面的接触面积很小。其中最诗意的建筑是位于因斯布鲁克的罗德帕克缆车道（Nordpark Cable Railway, 2004—2007，第166页至第169页）。复杂的钢筋网呈现出天篷华盖的造型，但却不止于此，涂有油漆的金属面板交织成一片熠熠生辉的白色云彩，而这片云彩被几个点拴在了扶梯和台阶上。这座建筑彰显着气场、敞开着情怀、定义并展现着风景，却丝毫没有限制自身的空间以及人们对它的阐释。在伦敦水上运动中心（London Aquatics Centre, 2005—2011，第180页至第185页）的造型中，这些膨胀的云彩在高空拢起一片空间，如同拱门一般，不费吹灰之力便可覆盖其下成千上万的观众和他们一旁的池塘。在生命的最后10年中，哈迪德开始将她那迂回曲折的流线型带入空中，她试图打开封闭的空间，试图将简单的杆状物变换为耸入云霄的蛇形、变换为正在绽放的花朵、变换为被门廊和巨大入口所打开的球状建筑。它们将与世隔绝、装有空调、四平八稳的建筑从地面上废除，使建筑不再与周围的环境相分离。

与在建筑规模方面的追求完全相反，哈迪德在构造基础材料时追求后客体艺术的现实感。她赋予家具以结构和形状上的自由，将它们从巨大而缠绕的大块构造中解脱出来，让它们延伸至空间深处，使它们的体积大小仅仅能容下一人。诸如位于土耳其伊斯坦布尔的卡塔尔·彭迪克总体规划（Masterplans for Kartal-Pendik, 2006年至今，第194页至第195页）、位于西班牙毕尔巴鄂的佐罗扎伊半岛总体规划（Zorrozaurre, 2003年至今，第150页至第151页）以及位于新加坡的一北总体规划（One-North, 2002，第131页），在这些城市总体规划设计中，她强调说所设计出的风景中其实包含着一系列的元素。其中，有一些是空间元素，有一些仅仅是风景中诸如街道般的起伏元素。她利用这些元素塑造出另外一种现实，使人们虽置身这种现实之中，却并不受限于结构、功能和场所的范围。在这凭借计算机技术所塑造出的毯状设计中，首先映入眼帘的只是起伏的风景，它们也出现在大规模的城市设计之中，卡塔尔·彭迪克总体规划便是一个典例。

风景随后变换为具有生命力的重写本，展现着那些凝滞在基础设施交汇处的瞬间。无论是迪拜的高塔，还是莱比锡城，或者罗马的管状构造，这些线条都垂直地延伸表达自己。它们延伸到风景之中，将空间囊括在线条之间。对于哈迪德和她的工作室人员来说，这些线条以高度凝缩的方式表达着建筑的作用，它们将我们所在的这个世界抽象化，将它敞开、重塑，然后定型为流动的形状。

在两座亚洲城市中，哈迪德展现了她的凝滞主题。它们分别是位于首尔的东大门设计广场（Dongdaemun Design Plaza, 2007—2014，第200页至第205页）以及位于北京的银河SOHO建筑（Galaxy SOHO, 2009—2012，第250页至第251页）。它们并不是简单的建筑物，也不具备彻底的都市特色，这些商业以及文化缩影建筑让人想起基本的动物造型：巨大的变形虫正盘绕在场地之上；水母正伸出触手将你环绕，它们将触手伸向街道和地铁；乌贼的圆脑袋占据了设计项目中的大部分。这些形象也许宣扬着参数作品的有机形态，但它们同样也向我们展现了许多艺术家和建筑师的发展倾向：他们在艺术生涯的后期，都更倾向于这种具备流动性的形状。比起哈迪德早期的作品，东大门设计广场更完美地将场地和建筑融为一体：它囊括了一处地铁站和一座公园，它成为了许多废墟和历史建筑碑的栖身地，它从博物馆、会议中心、商业区以及其他设计元素之间脱颖而出，它利用了这些设施的空间并赋予了它们更深刻的意义，它们一起将广场塑造为宏大而实用的城市结构。如银河SOHO建筑一般，哈迪德和她的工作团队用生涯中的第一座宏大结构，证明了他们可以以高度的控制力塑造出如此规模的结构，从而将广州大剧院那沉重的交叉点和摇摆的管道远远地甩在了身后。

哈迪德越来越热衷于寻找建筑中渐强的高潮以及凝滞在一瞬间里的狂喜。她所设计的阿布扎比表演艺术中心便是一个测试性的典型案例。该艺术中心面朝大海，它拔地而起囊括大规模的展现空间。此类建筑中，最为成功的是位于阿塞拜疆巴库的阿利耶夫文化中心（Heydar Aliyev Centre, 2007—2012，第216页至第217页）。这座建筑的线条类似于她早期作品中平行的条纹，但它以上上下下的陡峭坡度塑造出更具风情的空间，从而超越了早期作品中有限的空间。如果位于东京奥林匹克体育场中的水上运动中心已经建成，那么此类空间所暗示的寓意也将更为集中地体现在这座建筑的宏大风格之中。正在建设之中的卡塔尔足球体育场（Qatar Soccer Stadium）如今也将彰显出空间所暗示的寓意。

虽然哈迪德一直很喜欢这样的管状构造，然而，在她后期更为成功的许多作品中，她却拓展了多条管道的造型；她赋予了这些管道形状以棱角分明、戛然截断的末端造型，让它们发挥着观景台的作用。比如说，位于伦敦的伊芙琳·格蕾丝学院（Evelyn Grace Academy, 2006—2010, 第198页至第199页），以及高高耸立在阿尔卑斯山上的意大利梅斯纳尔山博物馆（Messner Mountain Museum Corones, 2013—2015, 第208页至第209页）。并非巧合，这样更为简洁的造型，似乎总是最频繁地出现在空间有限、预算拮据的情况下；它们不仅将迎面而来的新风景展现在饱满的流线型设计中，作为支撑建筑的结构，它们还让稠密的空间和形状变得更为复杂，它们容纳并编排着空间和形状，一路柔化着它们的棱角，并顺势打开它们流动的空间和造型——当项目预算或空间用尽的时候，它们便戛然而止，让我们从建筑的玻璃外捕捉到它们那被截断的末端，让我们欣赏那似乎是为激励我们探索而截断的形状吧。

作为哈迪德工作室的基本代表风格，斜面、管道和起伏在塔状建筑的正面展现着设计的风格。在建筑的底部，它们塑造着空间，将建筑的结构敞开并延伸至公共前厅；在建筑的顶部，它们营造出高耸入云的势头，而与此同时掩藏起自身的机械设备。当哈迪德再也不能在设计或背景中塑造出、引申出更多的空间时，她便经常利用场地所独有的特点（诸如角度或视角）或者结构的性质，打造出该设计所独有的特别之处。

在最后的10年里，哈迪德和她的团队的确发展出了一种替代性的设计模式。这就是岩石山，它也被称作人造山。通过这一设计，她为位于中国香港的赛马会创新楼（Jockey Club Innovation Tower, 2007—2014, 第220页至第223页）营造出了绝佳的效果。这座建筑周身布满条纹，拔地而起，以倾斜的角度伸向天空，建筑的弧度迂回曲折，偏离了垂直的混凝土地面，它的周身布满了管道。在建筑的内部，哈迪德分割这一空心的石核，塑造出空间并以桥梁、楼梯和电梯打破空间的狭隘，塑造出多样的弧度，进而创造出超乎她其他所有作品的崇高感。在位于比利时安特卫普的港口之家（Port House, 2009—2016, 第238页至第241页）中，她将许多人造几何图形打造成了一座岩石山，将其立于港口行政办公主楼之上。它将这座岩石山向远处推出一段距离，与此同时提醒着我们它那不可思议的重量。

虽然哈迪德过早地离开了我们，但是她却设计出了大量的作品，她通过自己的素描、绘画以及设计改变了建筑界。她是一位才女，她所设计的作品包括船只、水龙头、高楼大厦以及遍布全世界的城市规划。更加意义重大的是，她所设计的作品不仅易为人们所识别，而且还被人们争相模仿。她是全世界最伟大的建筑师之一。她从打造十分之一秒的爆破，成长到以结构容纳、赞美文化，并使之永存；计算机技术为她的设计带来了一场变革，她将计算机技术应用到从哈萨克斯坦到法国马赛的建筑工地上。蛇形结构蜿蜒而行、凸起的结构不断延伸、平面结构拔地而起、毯状结构则铺展开来。哈迪德一路向前，一生都在探索自己的艺术技法；她所设计的建筑仍然在不断地拔地而起，而她的影响，则一直接连不断地展现在我们的眼前。

注释

1. 瓦尔特·本杰明，"机械复制时期的艺术作品"，收于《启发》（Illuminations, 纽约：萧肯出版社，1969年）哈利·佐恩（Harry Zorn）译，第236页。

2. 阁楼展现了同样出色的现代主义空间。作为开放的实用工业空间，它消除了私人和公共设计与装饰之间的界限，并将我们从这二者的区别之中解放出来。这样的基本构造不仅出现在哈迪德的作品中，同样也出现在其他晚期现代主义的作品中，比如蓝天组的作品。关于阁楼的意义，在《蓝天组》（Coop Himmelblau, 伦敦：建筑评论出版社，1998年）这本书中做了更为详细的讲解。

3. 吉勒斯·德勒兹（Gilles Deleuze），《层叠：莱布尼兹和巴洛克》（The Fold: Leibniz and the Baroque, 明尼阿波里斯市：明尼苏达大学出版社，1993年）汤姆·康利（Tom Conley）译，第5页。

4. 同上，第34页至第35页。

5. 与作者交流得知，1997年12月14日。

6. 见曼纽尔·德·兰达（Manuel De Landa），"非有机形式"，收于《第六空间：融合》（Zone 6: Incorporations, 纽约：区域出版社，1992年），第128页至第167页。

7. 与作者交流得知，1997年12月16日。

伊芙琳·格蕾丝学院

建筑和项目

马列维奇的建构（MALEVICH'S TEKTONIK）
英国, 伦敦, 1976—1977年

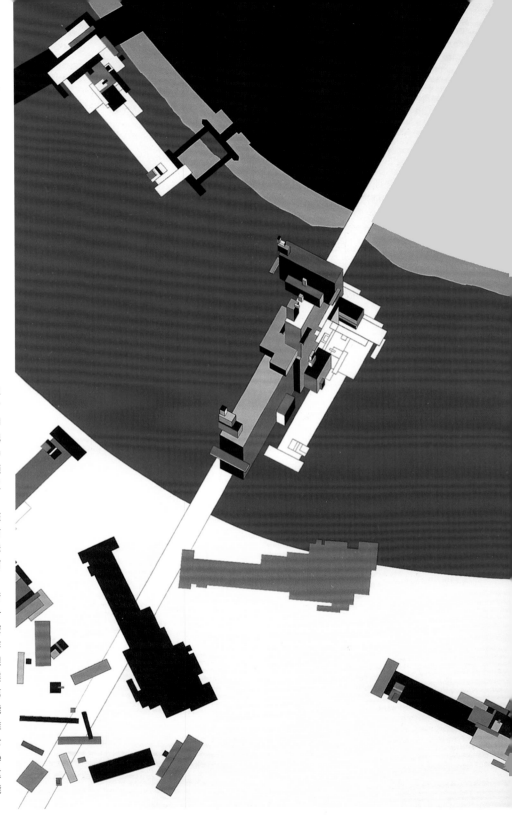

作为在伦敦建筑联盟的毕业设计，哈迪德为泰晤士河设计了一座亨格福德桥（Hungerford Bridge）。为了满足桥上一家客栈的设计需求，她试图探索设计中的"突变"元素。水平的"建构"顺应并利用了看似无序的至上主义形状，以满足设计和场地的需求。

在泰晤士河的南岸上，一座20世纪50年代的综合艺术馆中布满了野兽派造型，它们主导着南岸的艺术风情。这座桥将南岸与河流另一侧的19世纪风格相连接。这座建筑的14层结构系统地附着于马列维奇的建构之上，它们将一切可以感知的界限变换为新的空间可能性。这项设计与后期的作品遥相呼应：首先是位于纽约古根海姆博物馆中的伟大的乌托邦展览（The Great Utopia show，第63页），在这项展览中，哈迪德能够以具体的形状实现某些马列维奇的建构；其次是同样位于泰晤士河之上的宜居桥设计（Habitable Bridge，第84页至第85页），在这项设计中，她思索了发展混合功能的可能性。

在海牙市中心一座矩形的"堡垒"之中，坐落着一幢名为国会大厦的独立综合设施，大厦中所容纳的是荷兰议会与政府这两个政治体系。为了分开在政治上截然对立的这两个组织，便需要修建一处三角形的区域，以拓展调节议会争端的空间。因此，在设计这个项目时，不仅需要将设计区域置于已存在的结构之内，还需要在空间上给予议会一定的自主性。只要在议会大厦中搭造出一条分割线，便可以达到这样的目的。这条分割线是由两幢石板建筑实现的：首先是水平的元素（一道玻璃砖矮墙不仅具备多种功能，而且还可以充当有顶盖的讨论场所，方便议员们进行政治活动），其次是一座由椭圆厅所构成的小型摩天楼。一片集会空间将这两处建筑设计连接在一起，让公众和政府官员近距离接触；一道回廊游走在其中的一座石板建筑之中，方便人们在里面穿梭。

荷兰国会大厦拓展部分（DUTCH PARLIAMENT EXTENSION）
荷兰，海牙，1978—1979年
大都会建筑事务所：扎哈·哈迪德、雷姆·库哈斯、埃利亚·增西利斯

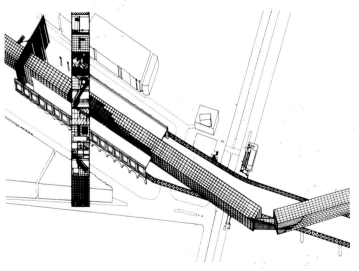

这座建筑是体现哈迪德早期思想和意图的建筑之一。在这座建筑中，她试图为19世纪末的建筑创立法则，以定义建筑在城市中所应扮演的角色。她尤其感兴趣的是历史与文化的来龙去脉。因此，她用两种方法探索了这座19世纪博物馆的原型：一种是通过详细地阐述都市所在位置的精确社会情境，另一种是通过展示建筑象征性的敏感性，而这种敏感性在当时语境主义建筑师的作品中往往是一种空缺。

19世纪博物馆（MUSEUM OF THE NINETEENTH CENTURY）
英国，伦敦，1977—1978年

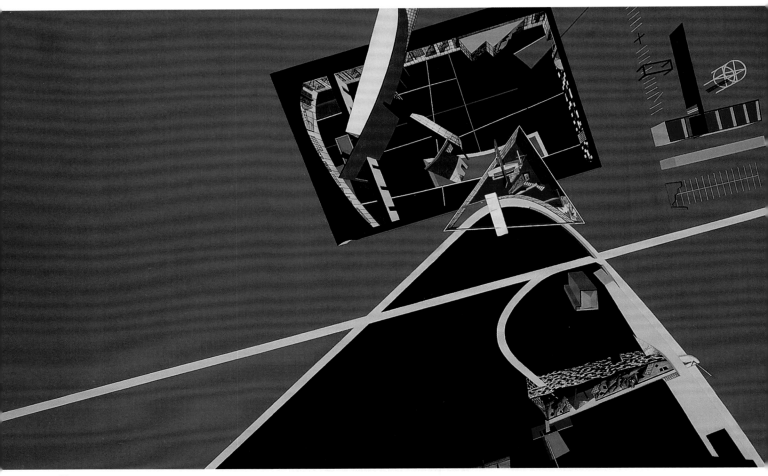

爱尔兰总理住宅（IRISH PRIME MINISTER'S RESIDENCE）

爱尔兰，都柏林，1979—1980年

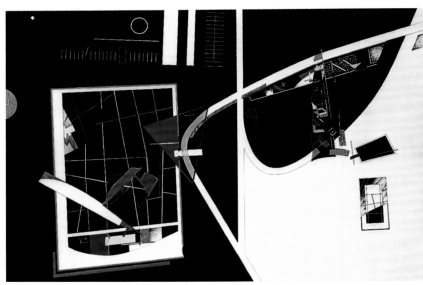

作为哈迪德的第一项主要设计，这一处新的住宅以及国务活动室是为一位爱尔兰总理所建造的。这项设计的目的是为了营造一种失重的状态，从而隐喻一种游离于公共生活压力之外的自由。一条马路和一条行人道连起了这两座建筑，但它们都需要保留各自的隐秘性。它们坐落在围墙围起的花园中，而花园是原本就存在的。若干间会客室隔开了接待客人的建筑和总理的住宅。在接待客人的建筑中，客人的房间分布在建筑内部的周边部位，与接待处和主卧套房相隔开来，而主卧套房则"悬浮于"花园的上空。

在一次比赛中，哈迪德为一家科技主题公园设计了规划方案和元素。这座公园坐落于巴黎中心最繁华的区域，其设计思路是在建筑工地平坦的地势之上，创造出悬浮的景观。旷野中绿色的高地构成了一座别具特色的花园：这座花园悬浮于此，而不是停滞于此。这些景观如同写在大地上的书法，机械系统使它们得以成型，将人文活动与无序自然一并控制在系统之中。野餐区、快餐店以及信息亭在它们自身的体系内运转，与狭长、单色的"行星带"形成鲜明的对比。与这项设计的未来主题相称的是，在这座公园中有一处"探索乐园"，这里汇集了公园的全部功能和景观。

拉维莱特公园（PARC DE LA VILLETTE）
法国，巴黎，1982—1983年

伊顿普雷斯广场38号的意大利领事馆的瓦斯爆炸为这一翻新工作提供了主要的灵感。这幢优雅的世纪之交联排房屋位于贝尔格莱维亚区的一条贫瘠的石灰路上。建筑内含三层，从概念上来讲，它们被转化成了三处垂直带。空间的交织意在带来特定的新颖感，丝绸以及卵石之类的建筑材料被应用于地面以及顶层之上，崭新的楼梯被安置在大厅以及餐饮区，将公共区域延伸至建筑的中层。

59号伊顿普雷斯公寓（59 EATON PLACE）
英国，伦敦，1981—1982年

匹克项目 (THE PEAK)
中国，香港，1982—1983年

　　这一度假胜地位于悬崖的顶端，至上主义地质（Suprematist geology）特色成了它显著的特点（建筑材料在垂直和水平两个方向紧密结合），它高高地超然于拥堵的城市之上。建筑物则如同利刃一般将场地一分为二，它们一边剥离了传统的构建原则，一边构建着新的原则；一边挑衅着自然，一边小心翼翼地避免破坏自然。

　　如同山体本身，建筑物也被分成了若干层，而每一层都具备自身的功能：第一层和第二层上布满了房间，第三层则因俱乐部的存在而别具特色，第二层与阁楼层之间存在着高达13米的空间，这片空间构成了新的建筑景观。在这片空间中，锻炼区、零食吧台、图书馆之类的元素如行星一般遍布其中。第四层与第五层上则分布着阁楼住宅。

　　象征着上流生活的顶点，也提供着一流的生活品质，匹克项目建筑以它的横梁和空间温柔地变换了以往的建筑风格，却为这千年不变的地貌带来了惊天动地的变化。

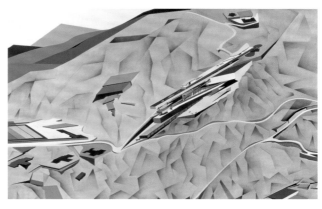
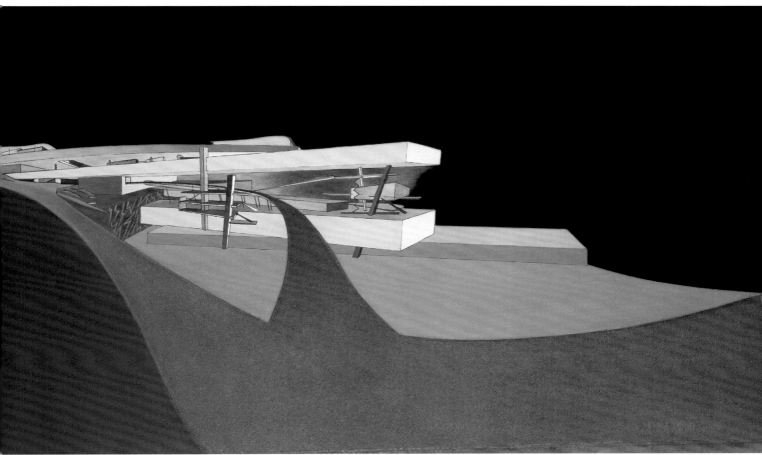

世界（89度）[THE WORLD (89 DEGREES)]
1983年

当哈迪德还是建筑联盟的一名学生时，她便开始在作品中探索未知的领域。这张绘图代表了她七年探索的顶峰。高速发展的科学技术和日新月异的生活方式，为这座建筑的设计提供了支撑，使它以全新的姿态脱颖而出，给人耳目一新的感觉。在新的世界环境中，她认为应当重启现代主义那未经时光见证便夭折的实践——并不是要复活它们的尸体，而是要揭开建筑新领域的面纱。在这张图片中，她凝缩并拓展了自己在过去七年里所做的设计。

梅尔伯瑞庭院（MELBURY COURT）
英国，伦敦，1985年

为了满足实际需要，英国人在二战后建立了许多窄小的公寓。这项设计的目的在于改造死板的小房间。在现有的公寓中，两面有弧度的墙上被贴上了镜面，它们偶尔相互重叠，围绕着中心处的光井营造出了畅然流动的空间。家具被置于过道或中心点，从而使生活区在空间以及功能层面更具灵活性。

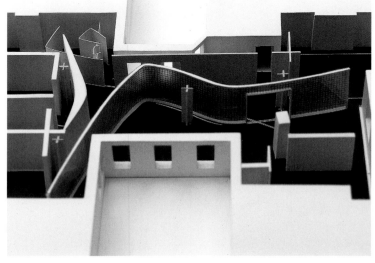

直到今日，试图重塑英国最著名广场的设计也依旧层出不穷。哈迪德的这一设计赞美了城市风景的动态可能性。它将公共区域延伸至专业办公区，从而拓展了现代建筑的领域，使之能够更好地提升都市生活的质量。公共平台、隔板工作室和塔状构造是这座建筑物的核心特色。塔状构造中分布着阁楼住宅，而在塔状构造之下，是隐秘的休息厅。一座购物广场拔地而起，商铺环绕在场地的周边，圈起了一片新的公共空间。随着大厦向上盘升，它便如同一处公共露台一般，俯瞰着其下成片的汽车。当你在广场上不断地变换着位置观赏这座大厦的时候，你会发现这些楼宇随着地面上碎片一般铺设的建筑发生变化，并使这些刺穿广场地表的建筑独树一帜。

特拉法加广场大厦（GRAND BUILDING,TRAFALGAR SQUARE）

英国，伦敦，1985年

哈尔金广场 (HALKIN PLACE)
英国，伦敦，1985年

从小面积的住房选址到更大范围的城市规划，这项设计从不同层面上考量了伦敦。哈迪德设计出了一片屋顶景色，将其置身于天空与下面的城市环境之间——有些屋顶是宜居的，而有些则不是 [这一创意预示着凤凰剧院 (La Fenice) 的设计理念 (第86页)]。在一座土地稀少、规划限制严格的大都市中，这片隆起的区域本身就是一片景观，其上的空间被垂直地分成了室内与室外的区域。在哈尔金广场的方案中，阁楼空间被设置在已有的屋顶和新建的屋顶之间。

东京装置作品展 (KYOTO INSTALLATIONS)
日本，京都，1985年

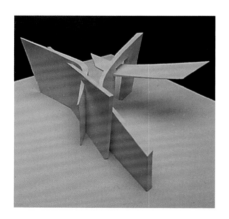

在这项装置作品展中，展现了某些零碎的概念，这些概念将在卡斯卡特道路24号 (24 Cathcart Road) 的设计中成型 (第29页)，正如大阪的愚人 (Osaka Folly, 第48页) 中的设计原则随后被应用到设计维特拉消防站中去 (第52页至第55页)。这项设计以新的方式在有限的环境中表达着空间，它用有弧度的墙壁包围或者是弯曲了空间，正如在梅尔柏瑞庭院 (Melbury Court, 第24页) 中那样，还用天篷装饰了入口。

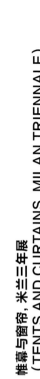

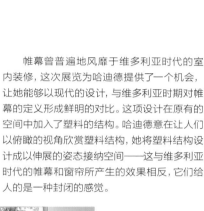

帷幕与窗帘，米兰三年展 (TENTS AND CURTAINS, MILAN TRIENNALE)
意大利，米兰，1985年

帷幕曾普遍地风靡于维多利亚时代的室内装修，这次展览为哈迪德提供了一个机会，让她能够以现代的设计，与维多利亚时期对帷幕的定义形成鲜明的对比。这项设计在原有的空间中加入了塑料的结构。哈迪德意在让人们以俯瞰的视角欣赏塑料结构，她将塑料结构设计成以伸展的姿态接纳空间——这与维多利亚时代的帷幕和窗帘所产生的效果相反，它们给人的是一种封闭的感觉。

卡斯卡特路24号 (24 CATHCART ROAD)
英国，伦敦，1985—1986年

国际主题住宅成全了哈迪德的第一次"至上主义地质"有形展，它实际上是哈迪德在59号伊顿普雷斯公寓（第21页）中所做探索的拓展。她所设计的比塔尔家具便出现在住宅里所配备的全套家具中。这些家具创造出了独立的动态空间，而不是像雕刻的物体一般被置于中性的容器之中。旋转和滑动，还有带储存功能的墙壁，这些可以动的门和橱柜使空间更加活跃。

在20世纪80年代晚期，人们非常热衷于在欧洲和美国的许多城市中改造升级海港区。两座工作室被设立在汉堡，以探索重塑这些地区的可能性。作为工作室的一部分，我们接到总体规划和设计城市历史海港区的任务，尤其是仓储城的前仓库区。我们要更新这一区域，并赋予它一系列的综合用途。

大片的、开阔的港湾，以及港湾与市中心相互交织的地理位置，为我们带来了一系列有趣的困扰。在汉堡的港湾道住宅小区（第44页至第45页）和科隆的莱茵港项目中（第66页至第67页），我们也曾以各种各样的方式思考相同的困扰。通过将城市的都市氛围带向港口，通过利用港口的独有空间——视野、开放性，以及常年存在的水域——我们寻找着新的格局与区域，我们希望通过重塑城市，不仅仅创造一种新的都市生活风格，还在城市的建筑物之间营造一种全新的动态。

汉堡港区（HAMBURG DOCKLANDS）
德国，汉堡，1986年

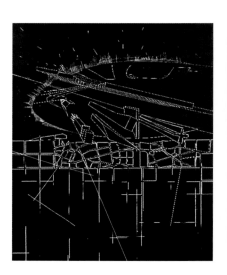

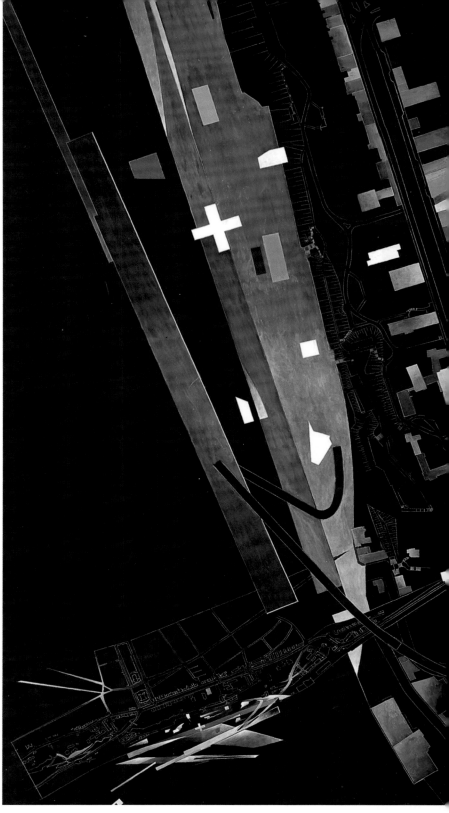

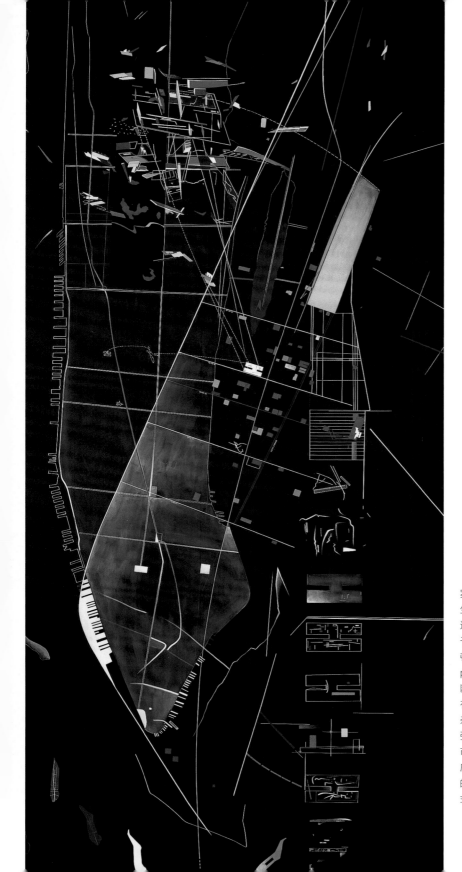

这幅草图所呈现的是一家酒店的重建方案。它勾勒出了我们在定义"酒店"和"都市生活"层面的可能性以及多样性，我们可以将这二者定义为某种有节制的释放。哈迪德对于这一层面的探索始于勒·柯布西耶在曼哈顿所设计的光辉城市项目（Ville Radieuse project），她认为这座城市从根本上错误地判断了纽约的城市环境。因为曼哈顿是一个具有多个层面的区域，城市的密度使其更为复杂，在这里建造的建筑应该具备一种凝滞的张力。勒·柯布西耶试图在设计中溶解这座城市，他仅仅想以一种现代主义的铺设来取代这座城市。而哈迪德则认为可以维持这座城市的强度，而无需消除将其拢聚在一起的棋盘式街道。

纽约，曼哈顿：一种新的书写艺术方案（NEW YORK, MANHATTAN: A NEW CALLIGRAPHY OF PLAN）
1986年

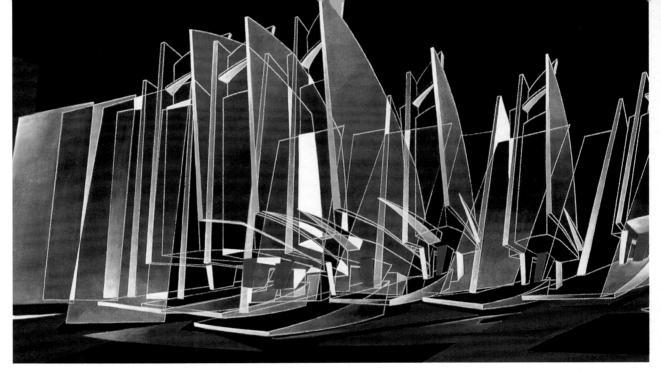

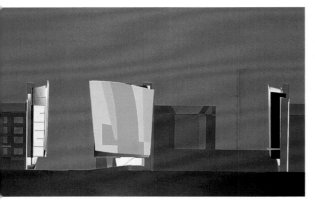

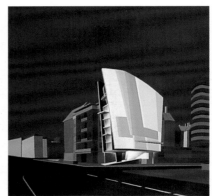

库达姆大街70号 (KURFÜRSTENDAMM 70)
德国，柏林，1986年

　　由于设计场地极其狭小有限（2.7米×16米），这座建筑便被赋予了压缩的"夹层"结构。它由一系列的平面、空间和功能构成。夹层中的水平平面是建筑平面图的基础，它在办公场所中实现了循环和运动的分离。在垂直方向上，空间分层实现了底层平面（这里设有该建筑的公共入口）与其上悬臂式建筑（用于容纳多间办公室以及一间顶端双层办公室）之间的分离。大厅和入口均高于地面，它们通过斜坡与地面相连，从而使得该设计不拘泥于地面，并以这样一种方式向德国至上主义者致敬。上部的结构与一道新的后墙相分离，斜坡之上的空隙将我们的视线引向主入口。

　　平面缓缓弯曲，并不断地向着角落处延展，因此，最高处的地面面积也最大，从而营造一种动态的效果，以拒绝普通办公建筑的千篇一律。长长的街道立面由透明的表面所构成——如同一张铝制构造网从顶部悬吊下来——从而形成了一个发光的玻璃盒，透视着建筑内部的活动。

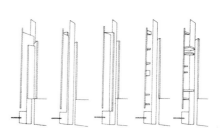

剖面图

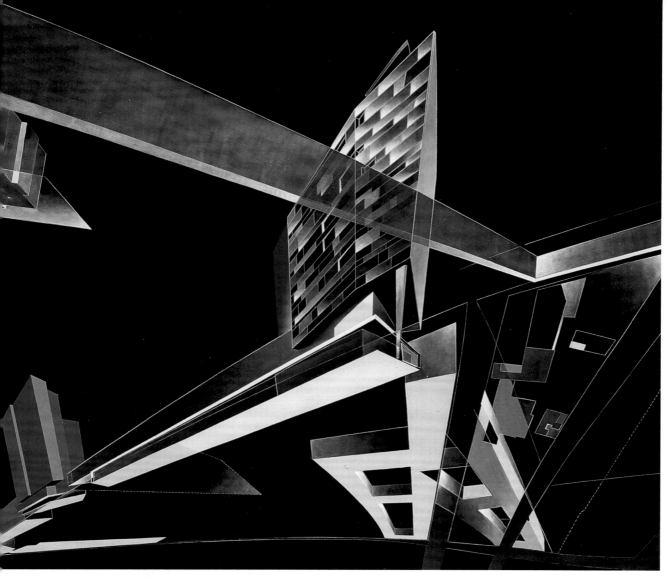

在介绍这项设计之前，需要强调两个基本的问题：一个是柏林国际建筑展（Internationale Bauausstellung）的填实和修复策略，另一个是有关社会住房的严格建设规则。后者有悖于现代开敞式平面布置设计。尽管存在着以上的这些限制，周围的建筑却也广泛地展示出不同时期的不同类型。所以，尽管规则规定此处的新建筑必须平均高达五层，在这留有余地的建筑环境中，天衣无缝地嵌入一座另类建筑不是不可能的。

一座高达八层的高塔被搭建在了一座长的三层高的长建筑末端，从而解决了以上所提及的规划问题。高建筑的底层设有商铺，而其上则为常规住宅；天台上设有屋顶花园以及儿童游乐园。如同雕刻般的建筑表面覆盖着一层经过阳极化处理的金属片，建筑内部的每一层楼都设有三间楔形的阁楼。

IBA-2号住宅（IBA-BLOCK 2）
德国，柏林，1986—1993年

麻布十番和富谷的合并平面图与剖面图

麻布十番（AZABU–JYUBAN）
日本，东京，1986年

库达姆大街项目（第32页）展示了哈迪德在释放空间方面的巨大潜力。在东京，大多数建筑场地都超越了空间的承受极限，许多建筑仅仅是加剧了城市那令人窒息的拥堵。

这座建筑拔地而起，嵌入了风景之中。它位于六本木地区附近一条混乱的建筑带中，这座建筑加剧了狭窄建筑场地的拥堵。古朴的玻璃结构介于高大的金属墙和加固的混凝土墙之间，其上镶嵌着宝石色彩的窗户。在这两面墙之间，是另外两面玻璃幕墙：一面是蓝色的玻璃，另一面是透明的玻璃。它们向外倾斜，伸向阳台上的护栏。在建筑内部，你能够立即捕捉到空间被释放后的效果，三层重叠的入口明显地诠释了这一效果。一架垂直的楼梯从建筑的中心处一直延伸到顶部，直到风格夸张的阳台。

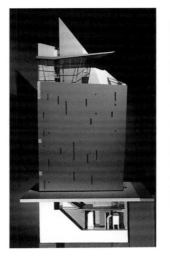

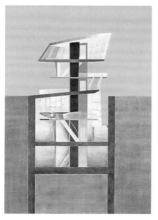
剖面图

　　这幢小型多功能设计的建筑位于杂乱的居住区，它在诸多方面都与麻布十番相关联（第34页），但我们此处所提及的创作理念刚好与之相反。楼梯和平台以螺旋的态势，将一系列悬浮的平行空间与垂直元素交织在一起。建筑师在这幢建筑的内部营造出了空荡荡的效果，而并非像在麻布十番中那样，将空荡荡的感觉排挤于建筑之外。设计的中心是一座精致的玻璃凉亭，它三面敞开，悬浮在开敞的地面上。这座建筑的大部分位于有弧度的地面层之下，地面层则被拉起来反扣在地面上，而一面高大的玻璃墙使得阳光能够照射到较低的空间里。这座建筑中豁达的设计比例非常灵活，使人们可以在这里进行零售以及办公等活动。

　　在这样一座拥挤的城市中，光线和空气是宝贵的配置，我们需要将空间从有限的场地中释放，将光线和空气投放到城市环境之中。

富谷（TOMIGAYA）日本，东京，1986年

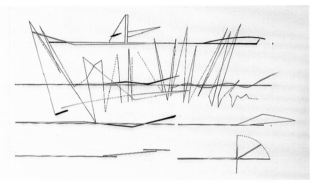

在该项目的几个饶有趣味的设计方面,首先是相对年轻、进步的市政当局希望将自己的标志融入洛杉矶复杂的布局中去,其次便是这里有着丰富的创造性资源——内部装修设计师以及绘图师高度聚集在美国的西好莱坞。西萨·佩里(César Pelli)的太平洋设计中心就在距离这里不远的地方,它被人们称作"蓝鲸",是这座城市最容易识别的建筑标志之一。

这里的建筑环境相对开放,哈迪德在这里初步探索了将建筑塑造为风景的方式。这里平坦的地势使她能够将施工场地想象成几何地势,她曾在拉维莱特公园中展示这一手法(第21页)。在这片几何图形的画布上,她所绘制的建筑物漂浮着与彼此互动,展现出唯有在开阔空间中才能实现的效果。

西好莱坞市府大厦(WEST HOLLYWOOD CIVIC CENTER)
美国,洛杉矶,1987年

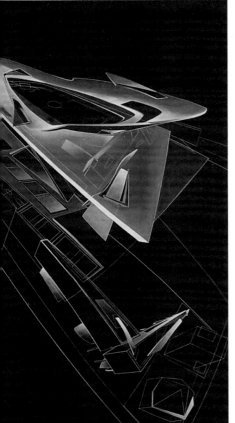

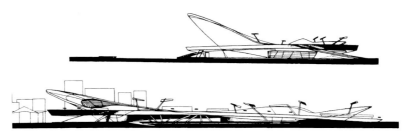

立面图

阿尔瓦赫达体育中心（AL WAHDA SPORTS CENTRE）

阿拉伯联合酋长国，阿布扎比，1988年

这项为稠密的城市所做的设计，与为柏林以及东京所做的设计形成了鲜明的对比（第32页至第35页）。这座体育综合设施的布置使我们能够深入环境中去，而并非将环境变得更为拥堵。因此，这座建筑结构作为一处大规模的风景，缓解了环境中的压力。它映衬并介入到环境的轮廓中去。在某种程度上，在这项设计中，哈迪德对风景进行了初期的探索。它包含着三种主要的元素：裙房，作为一处悬浮的停车场，它为体育场以及看台提供了入口；新的透视地平面，从与路面平起的高度向上伏起，一直延伸到裙房之下；体育场，坐落在变换的透视地平面以及裙房之上，具备各种功能，提供了各种各样的就座设施。

大都市 (METROPOLIS)
英国, 伦敦, 1988年

这张图片上咆哮的红色意在表示恼怒, 恼怒于这座城市中张牙舞爪的混乱。它被用于伦敦当代艺术中心展览中, 该展览意在探索城市的不同侧面。一方面, 这张绘图将伦敦展现为一片村庄大杂烩。但我们并没有进一步加剧这座城市布局的融合——它已经以这种布局发展了数世纪之久——我们以多中心的方式贯穿这座城市, 也就是说, 不同的城市中心凝滞在不同的焦点上。在这一背景下, 红色象征着伦敦的火焰, 新的机制和中心需要被创造出来, 以代替筋疲力尽、负荷过度的中心。

柏林墙被拆于1989年，在这之前，我们曾被邀请思考这座城市的未来。作为从梅林广场到弗里德里希大街和从勃兰登堡门到亚历山大广场总体规划的一部分，柏林墙的倒塌为城市更新提供了新的可能。我们考虑过扩建和修复城市，小到廊道的规划，大到"城墙区"的建筑设计。

　　我们关注的焦点是亚历山大广场。因为它代表了为数不多的几次尝试，我们尝试超越19世纪独特的都市特性，我们打算使其脱离千篇一律的商业发展，与那条脆弱的线，那道曾用于划分柏林的线形成鲜明的对比。关于新释放地域的发展，一系列的绘图提供了诸多可能性。走廊设计蜿蜒在风景之中，新的几何图形入驻先前"死寂的区域"（见下图），它们有时以直线的方式包围广场，稍稍偏离了既存的秩序。在我们的设想中，城墙区将成为一座带状公园。这里不曾有过一面混凝土带状墙，也不曾有过一处禁行区，我们可以在这里设计一座带状公园，并以建筑物将其装点。

柏林2000（BERLIN 2000）
德国，柏林，1988年

维多利亚城市区域（VICTORIA CITY AREAL）
德国，柏林，1988年

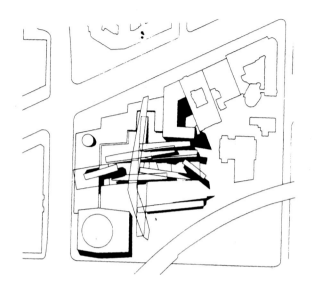

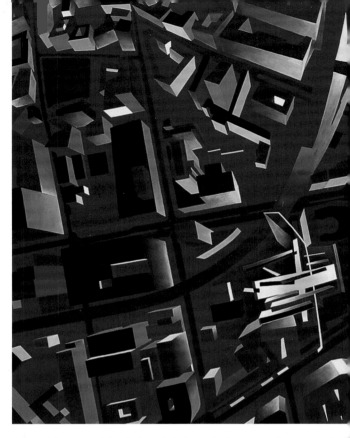

在推倒柏林墙之前，作为城市中的一座孤岛，这一地区缩影了整个城市的状态。虽然库达姆大街位于城市的主要轴线上，但它却被彻底地围了起来，完全不让人靠近。若要在这样严加防范的环境中建起一座建筑物，则意味着我们要在平面层次上提升城市的密度。因此，建筑场地被分割成了新的空中走廊，三处独立的区域具备各自的主要功能：购物设施、办公室和酒店。

因为十字形的建筑场地将成为数条主要线路的焦点——街道和地铁线路——一处新的购物区则被置于梯形区域的中心点上，其周围层层向外分布着商铺。在这一圈起的空间中，地面上被铺以玻璃，其下布置着公共设施，这些设施中包括多家商店、一家宾馆大厅、一座多功能礼堂、一家会议中心以及一家酒店。在其上，一间间办公室的横梁交织成一个系统，其规模还可以再扩展——每一根横梁都标示着一家企业独特的身份——它们被附于购物设施之上。在横梁系统之上，悬浮着一面弯板，它托起了整座酒店。

新巴塞罗那（A NEW BARCELONA）

西班牙，巴塞罗那，1989年

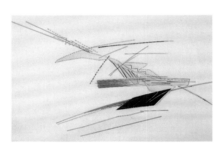

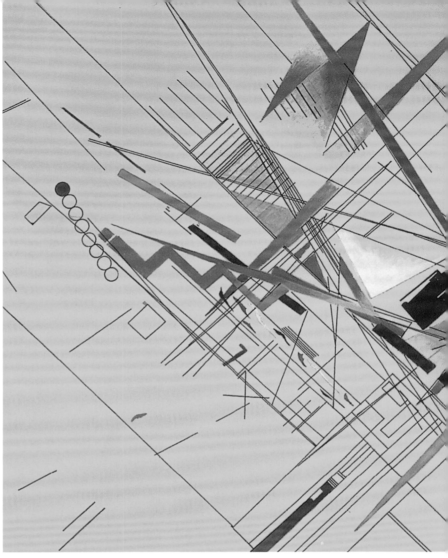

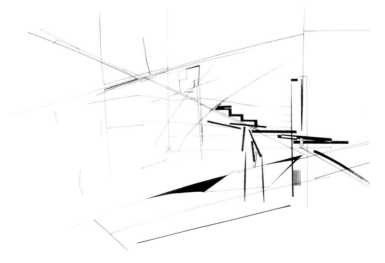

伊尔德方斯·塞尔达（Ildefons Cerdà）曾规划巴塞罗那的19世纪城市扩展建设，它所做的对角轴成为了我们重建这座城市的关键性元素。我们将对角轴轻轻地扭成了歪斜的、交织的碎片，从而构建出新的城市形状。这一区域的地势在城市氛围中蔓延开来，一条为当地量身定制的"景观变换带"频频与其相交织——不规则的部分是村庄，网格状的是住宅区，而带状的则是铁路和码头——它激活了整座城市，使邻里间的街道活动多样化。

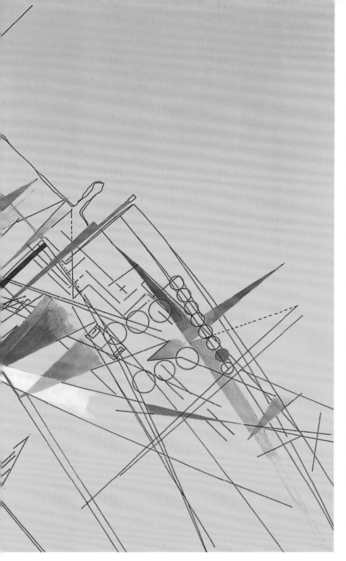

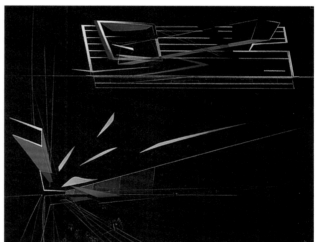

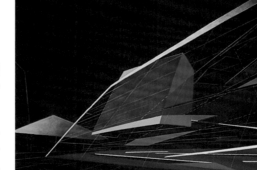

与在麻布十番（第34页）以及富谷（第35页）中所运用的原则相似，这项设计意在对抗东京的拥挤。其中心是一个空的类似于玻璃容器的设计，在其四周遍布着规模较小的中空设施，它罩住了这座建筑物中的文化区以及会议区。

在紧密交织的房间所构成的迷宫中——正如庞贝城那样——各种各样的隔板将聚集在一起的会议区分隔开。在地面层的玻璃地板上，有一道切开的裂缝，透过它可以看见下面的空间。建筑上层设有展览区、工作室、酒店以及公共区，屋顶上建有一座景观花园，一条沿对角线切割的裂缝使光线和空气能够进入较低的楼层。

东京国际会议中心（TOKYO FORUM）
日本，东京，1989年

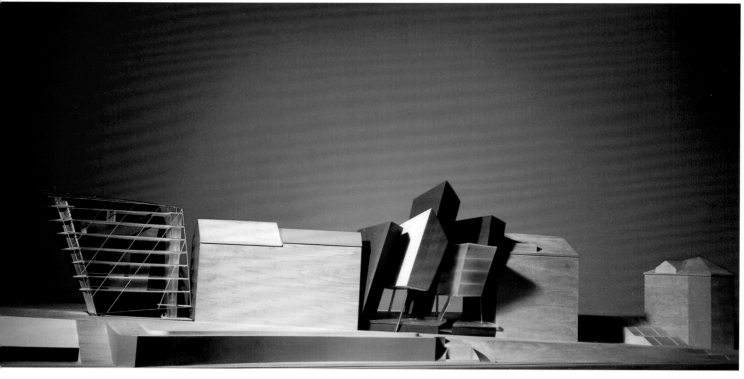

港湾道住宅小区（HAFENSTRASSE DEVELOPMENT）
德国，汉堡，1989年

在古老的港口街中，布满了四层或五层高的旧式房子。那里有两块场地——或者不如说是两道缝隙——有待开发。这里有许多相互平行的建筑带，由街道以及排屋构成（小公园、新街道以及路堤，构成了别样的元素），它们一直向下延伸到易北河边。我们的任务是将建筑带相互贯通起来，并将路堤地带改造成娱乐区。

其中一处场地位于尖锐的拐角处。一座前倾的板楼建筑稳稳地伫立在地面上，它面向河流所在的地带。这座建筑被分成了商业层和住宅层，在前两层上设有公共的空间。玻璃幕墙上装有滑动装置，所以每一层的部分空间都可以被改造成户外阳台。面向河流的立面上造有不断延伸的幕墙，它向上延展，弯折而上成为阁楼的屋顶。

第二处场地位于一条19世纪街区的缝隙中。我们预想出许多紧紧挨在一起的平板，尽管它们很密集，但是却具备一定的透明度。当你在这座建筑中穿梭时，你会看到空间在建筑的缝隙之间敞开又闭合，它的立面并不是一成不变的。建筑底层是零售空间，住宅区域位于上层空间。这项建筑设计中的诸多方面的建筑原则被用于设计杜塞尔多夫的佐霍夫媒体公园（第50页至第51页）。

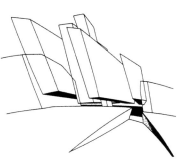

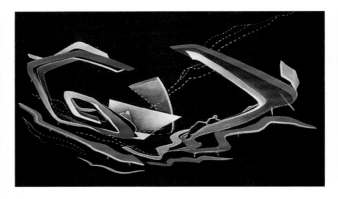

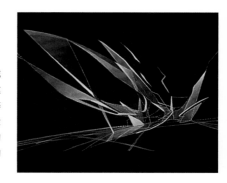

季候风餐厅 (MONSOON RESTAURANT)
日本, 札幌, 1989—1990年

在我们所设计的这座双层建筑中, 设有专门的就餐厅和休息室, 但我们想再营造出一种截然相反的氛围。于是两个互不相同却连为一体的世界诞生了: 冰与火的对峙。札幌的季节性冰建筑为我们提供了灵感, 这座建筑的底层为冰冷的灰色调, 材质为玻璃和金属。桌子由锋利的冰碎片构成, 凸起的楼板平面像冰山一般悬浮在整个空间中。在冰屋的上方悬挂着一个火炉, 它被涂成了灼热的红色、明亮的黄色以及旺盛的橙色。吧台之上的螺旋状构造向上延伸, 刺穿了底层的天花板, 它卷曲着向高层的圆顶下部伸去, 像一团火焰构成的龙卷风, 正呼啸着破出压力重重的容器。就餐室和休息室中提供不同类型的座位, 像可以移动的托盘式座椅和嵌入式的沙发。

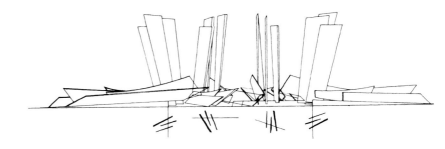

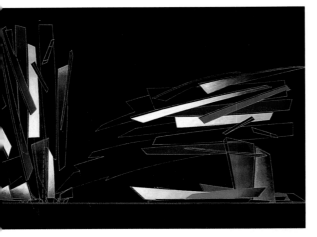

平面图和立面图

大阪的愚人，1990年世博会（OSAKA FOLLY, EXPO 1990）
日本，大阪，1989—1990年

　　我们的施工地点位于大阪的世界博览会，它坐落在一个露天的广场上，数条道路在这里交织。我们设计出了一系列压缩和混合的元素，在展现风景的同时也折射出行人的踪迹。远处，两面垂直伸出的木板象征着迎接游客的"愚人"，而在近处，平行分布的木板则定义了该结构的范围，并创造出了一系列的峡谷。五条长度不一的斜坡与飞伸而出的木板形成了鲜明的对比，它们沿着地面的木板延展开来。这些动态的水平元素与垂直元素出人意料地交织在一起，产生了许多的小幽谷。当游客们热情洋溢地散步和赏景时，也可以在幽谷中小憩片刻。

　　设计中布满了一道道蜿蜒的墙壁。我们可以将大阪的设计看作是维特拉消防站的半试验品（第52页至第55页）。

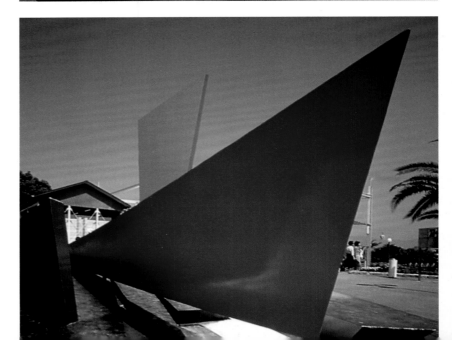

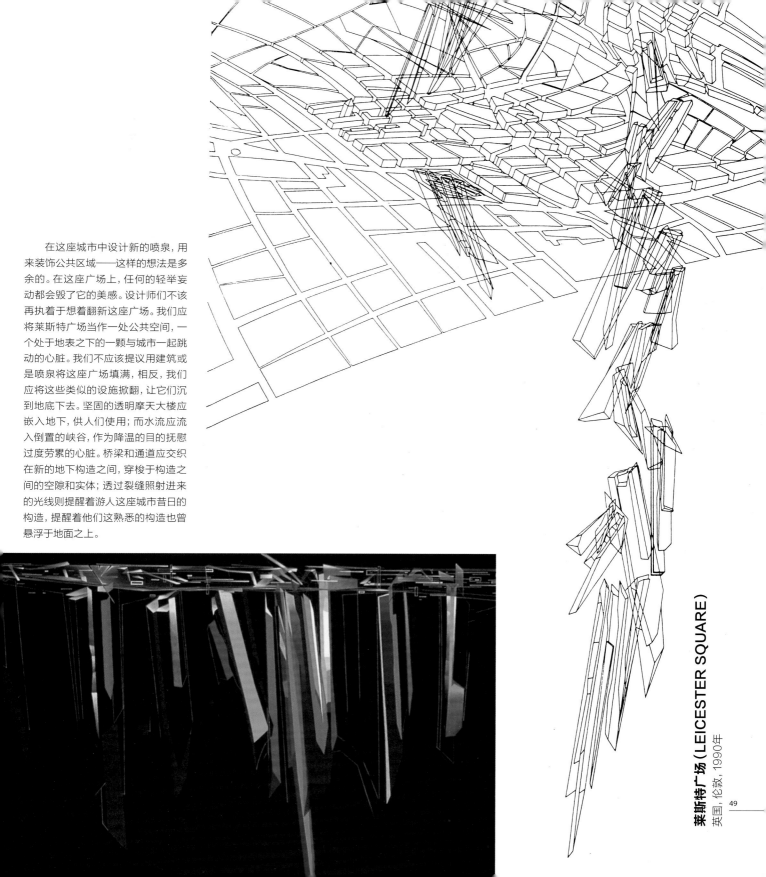

在这座城市中设计新的喷泉,用来装饰公共区域——这样的想法是多余的。在这座广场上,任何的轻举妄动都会毁了它的美感。设计师们不该再执着于想着翻新这座广场。我们应将莱斯特广场当作一处公共空间,一个处于地表之下的一颗与城市一起跳动的心脏。我们不应该提议用建筑或是喷泉将这座广场填满,相反,我们应将这些类似的设施掀翻,让它们沉到地底下去。坚固的透明摩天大楼应嵌入地下,供人们使用;而水流应流入倒置的峡谷,作为降温的目的抚慰过度劳累的心脏。桥梁和通道应交织在新的地下构造之间,穿梭于构造之间的空隙和实体;透过裂缝照射进来的光线则提醒着游人这座城市昔日的构造,提醒着他们这熟悉的构造也曾悬浮于地面之上。

莱斯特广场(LEICESTER SQUARE)
英国,伦敦,1990年

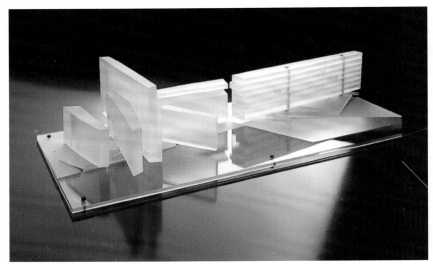

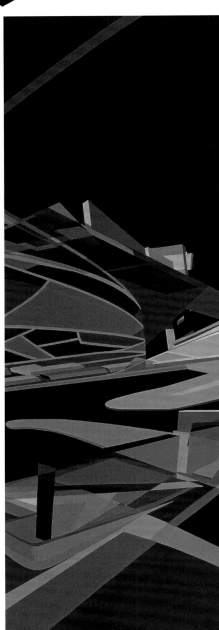

佐霍夫3媒体公园（ZOLLHOF 3 MEDIA PARK）

德国，杜塞尔多夫，1989—1993年

为了将杜塞尔多夫著名的港口打造成娱乐区（包括广告商办公室、散布在商铺之间的工作室、酒店以及休闲场所），我们设计了一处人工景观。它面朝河流，成为了水上活动和功能的延伸区域。如同一道屏障，一道长90米的建筑保护着这一景观，它不仅包围了办公室，而且还将交通噪声隔离在外。一架巨大的金属三角造型从河岸所在的方向切入施工地点，它刺破墙壁形成了入口处的斜坡。斜坡旁的接地平面上裂开了一道缝隙，朝向南方，展现着技术工作室、商铺和酒店。一道提供支撑的墙嵌入地下，地表之上的部分则被翻折下来，形成了一座可容纳320人的电影院。

在墙壁面向大街的一面，细小的线状切口嵌入墙壁之内。而在朝向河流的那一面，每一层上不同深度的悬臂则展示着建筑的分层。这座镶有玻璃的"指状"建筑是由许多片碎厚板拼接起来的，它垂直地立于街道上，一片片的玻璃仿佛挣脱了墙体的束缚而获得了自由。在厚板相接的地方，一片空间被镂刻出来，以容纳会议室和展览区。前厅是一座极简主义式的房子，它位于墙壁和指状建筑的相接处，被许多足状和厚重的三角状结构所环绕。在这里，你可以看到街道和海岸风景交织在一起。一架丝带一般的楼梯一直延伸到会议室，它穿过了沉重的厚板，悬浮在建筑的上方。

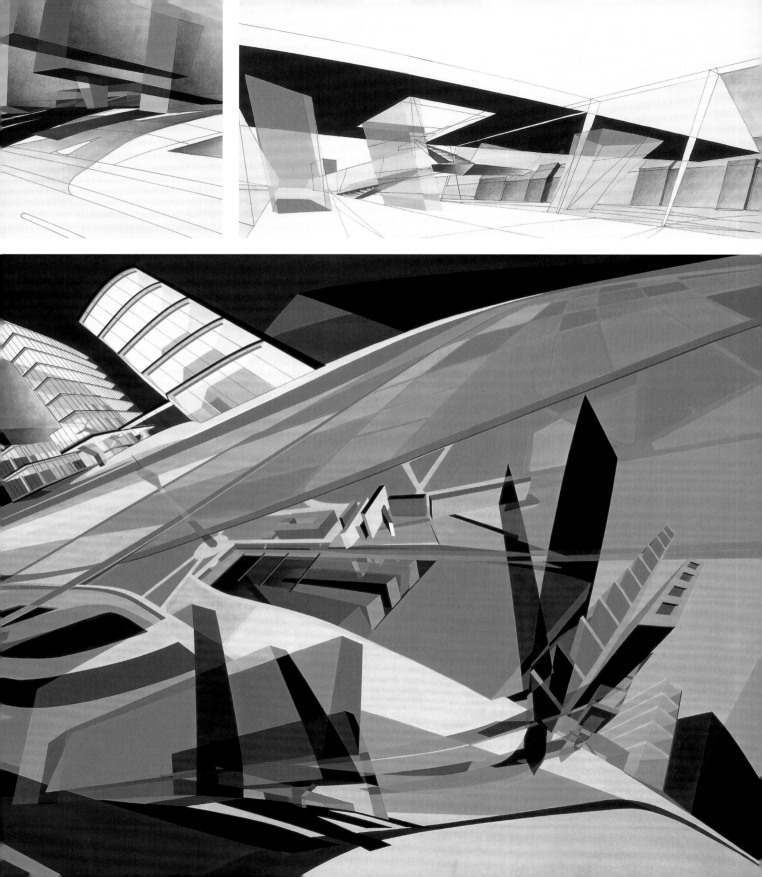

起初,我们接到的任务只是在维特拉最大的工厂综合设施东南区修建一座消防站,但我们将简单的设计拓展成了围墙、自行车棚以及运动场。这座新建筑物定义了空间,它所具备的功能展现了设计的构思:建造一排多层次的墙,消防站就坐落在墙与墙之间,一面面墙则按照功能的需要被打穿、倾斜或是截断。

现在消防站被用于举办活动和展览。只有用与建筑相垂直的视角审视时,才能看到其内部的装修。当游客漫步在消防站的空间中时,他们不时可以瞥见巨大的红色汽车,轮胎的痕迹留在了沥青路面上——如同舞者舞过的轨迹。整座建筑表现着凝滞在一瞬间的动态。它展示着蓄势待发的紧张,以及每时每刻都能够一跃而起的潜力。巨大的拉门也是可移动的墙壁,一面面墙壁迎着彼此被拉来拉去。

整幢建筑都是由裸露的、加固的混凝土构成,它们的边缘显得非常锋利,这一点非常引人注目。屋顶没有边饰,也没有镀层,因为它们会破坏菱形造型的简洁。这座建筑中没有多余的细节,而无框玻璃、滑动门以及照明方案——这一切则保证了消防车能够快速而精准地穿梭在消防站之中。

维特拉消防站(VITRA FIRE STATION)
德国,莱茵河畔魏尔,1990—1994年

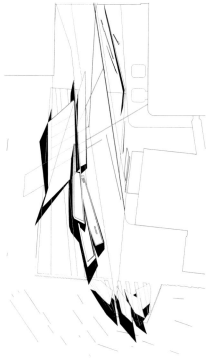

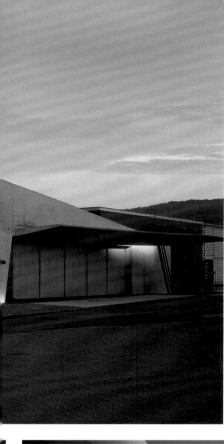

音乐影像展示馆 (MUSIC-VIDEO PAVILION)
荷兰，格罗宁根，1990年

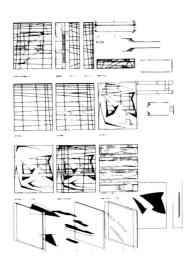

这项设计的思路是在这座城市中最具挑战性的地点建立一处娱乐场所，它位于鱼市区的阿科特建筑（Aa-kerk）和克伦博尔斯（Korenbeurs）建筑之间。正如纽约火烧岛上的"监控"房那样——面朝大海的隔板房，镶有巨大的平板玻璃，只在夜间才能看清的高级内部装修——这座展示馆的设计理念在于向世界提供一个窗口，使人们能够在视频影像中走来走去，成为表演中的一部分。

在两面墙壁之间相隔一米的地方，装饰物凸出至光滑的外壳处。影像从上层的层面投射到中层层面之上，进而投射到半透明的面板上（面板嵌于镶有玻璃的立面上），最后投落在其下起伏地面上的装饰上。为了制造出简短的讯息和视频，我们投入了大量的资金和人力，然而这些视频本身却不够完美。我们希望的是能够为导演、演员以及制片人提供一个框架，让他们依此进行尝试。

备选方案：平面图、立面图、剖面图和结构图

酒店与住宅综合设施（HOTEL AND RESIDENTIAL COMPLEX）

阿拉伯联合酋长国，阿布扎比，1990年

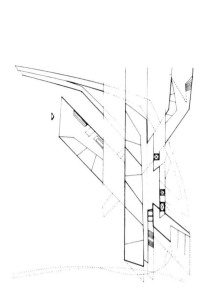
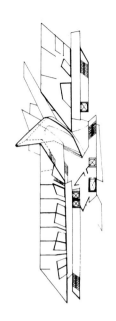
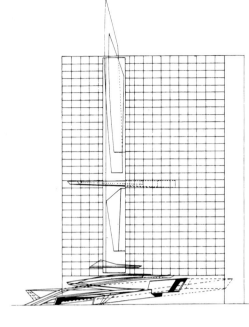

像美国的许多城市一样，阿布扎比的街道布局为棋盘式。棋盘式布局是一种均衡的结构，作为基础性依托，它使独特的建筑得以"林立"在城市中。当我们为市中心最佳地理位置的一座酒店综合设施设计方案时，我们将这面水平的城市棋盘掀了起来，从而形成了一个直立的平面；也就是说，公寓和酒店房间组成的平面成了酒店配套空间的背景：会议室、酒店以及健身俱乐部。在平面一分为二的地方，这些配套空间便悬浮在"垂直的院落空间"之中。在地面层上，一道四层楼高的建筑板直穿过这片场地，它穿透了平面，为上层的购物中心以及下层的办公空间搭建起空间。在四层楼高的横梁上向下俯视，你会看到酒店的大厅。一条弧形的斜坡从空地的一角俯冲至大厅门前，它绕过平面伸入垂直的院落之内。

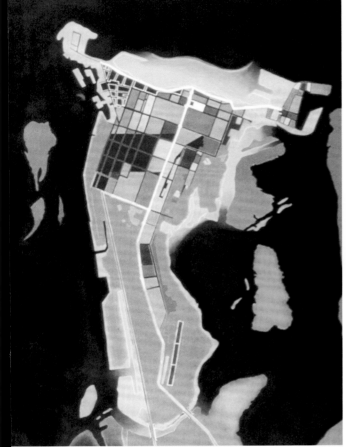

一位德国木材加工商让我们设计一个展示台,供他们在一年两次的室内装饰贸易展上展示产品。我们在设计展台时,起初的意图是要创造出独立的多层空间。以庞大而乏味的商业展览为背景,我们意在营造一种效果——将与世隔绝的景观处于密闭的结构之中。在展馆内部,一条中心道路如同一条满是分叉的树干,将人们引向产品展示台和周围的环境之中。

1991年德国室内装饰贸易展(INTERZUM 91)

德国,格鲁伦多夫,1990年

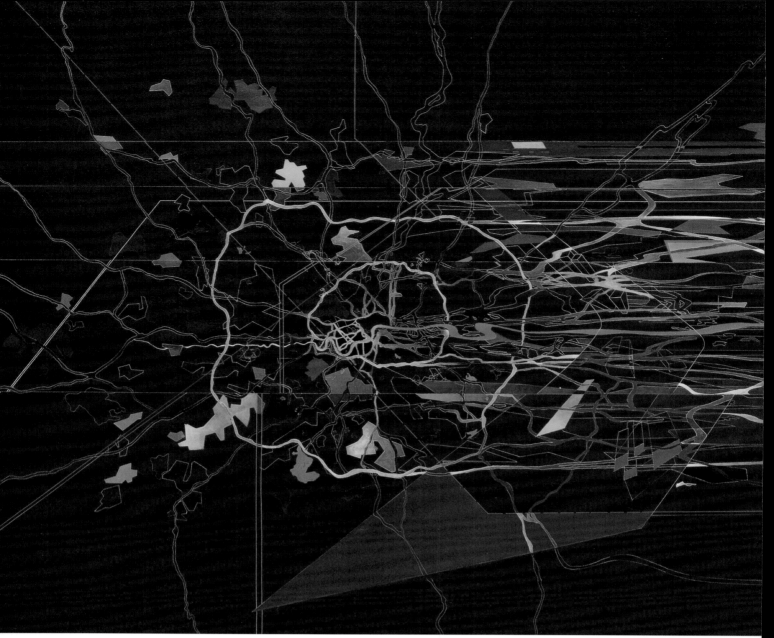

伦敦 2066 (LONDON 2066)
英国，伦敦，1991年

在英国《时尚》杂志所展出的这幅图片中，我们进一步探索了风景在伦敦都市中的性质。在设计特拉法加广场大厦（第25页至第27页）、大都市（第38页）以及莱斯特广场（第49页）时，我们便已经开始了这样的探索。这项设计以一幅图画的形式，展示了我们在概略与形象化层面，对于城市最为激进的改变——同样，它也应受到同等激进的评论。我们研究了露天空间、铁路、道路、水路、航线以及区布局，并重新构造了整个设计方案。笔尖从伦敦的西部开始画起，笔迹交织，一直画到了伦敦的东部。所绘线条与空中以及地面的分区线相交，我们认为在相交处可以建造建筑物。因为在这项规划中，正是在垂直结构与地面相交的地方，公共活动最为稠密和活跃。

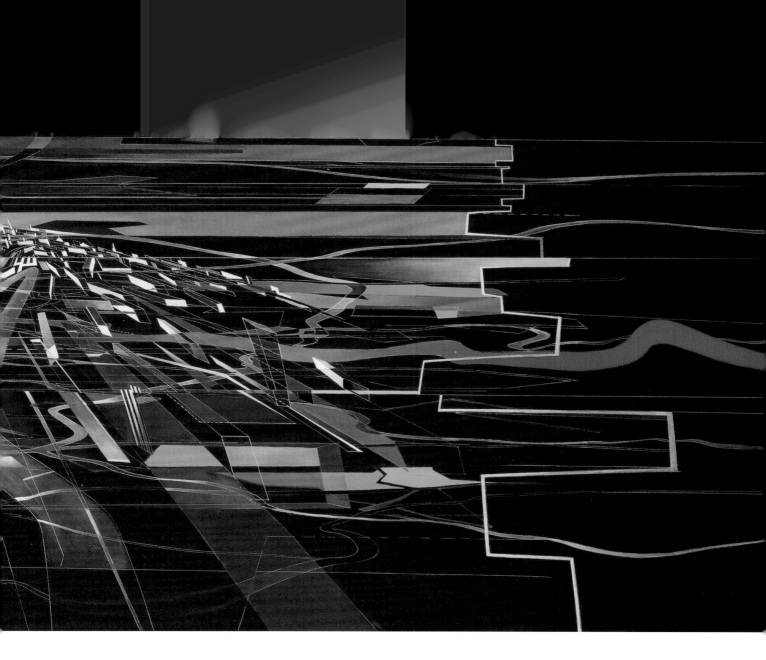

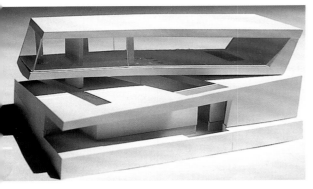

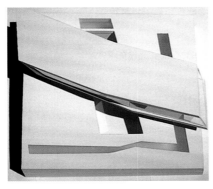
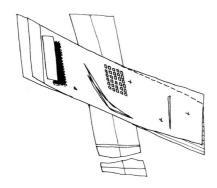
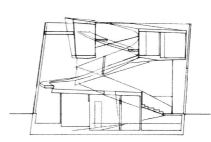
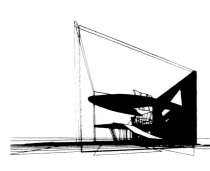

"十字"别墅草图,螺旋别墅剖面图

海牙别墅(THE HAGUE VILLAS)
荷兰,海牙,1991年

在海牙住宅设计节上,我们展出了两项设计作品。作为八幢别墅"区"的一部分,它们坐落在城市的郊区。说到别墅的设计,我们将其内部空间的常规构造加以抽象化,让空间与外界形成互动,从而营造出意想不到的感觉。第一项设计为"十字"别墅,其底层为一座裙房,两根"建筑板"在这里交错着支起了住宅的大部分面积。较低的建筑板切入裙房层,营造出"负空间",从而形成了一座庭院。上层建筑板营造出了"正空间",露天的居住空间与工作室空间则位于裙房之上并覆盖了庭院。另一座螺旋别墅本质上是一个立方体,一块楼板从其底层入口处螺旋上升至居住空间,然后抵达一间工作室,在某些地方还刺穿了外墙板,玻璃立面随着地面的盘旋上升,形成了螺旋的造型,从而交替展现着实体式、百叶式、半透明式以及最终透明式的风格。外墙板的空隙与内部的螺旋造型之间是居住空间和缝隙,景观和通道相互交错、相互作用,给人耳目一新的感觉。

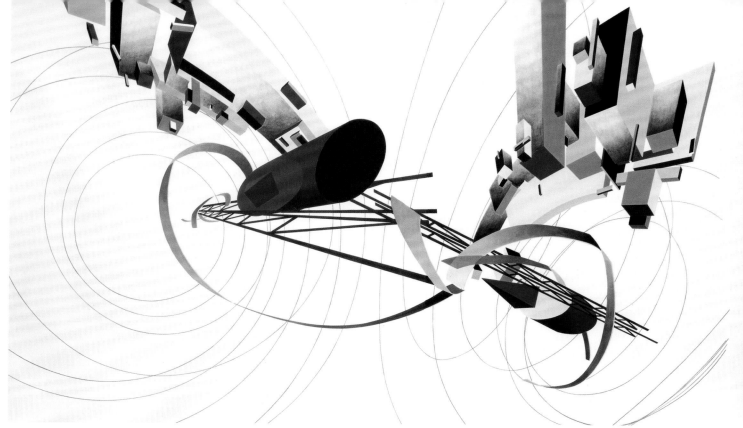

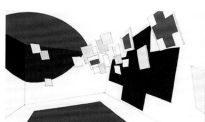

这项设计被应用于俄罗斯至上主义以及构成主义展览。这次在纽约古根海姆博物馆所举行的展览为哈迪德提供了一个机会,让她能够重拾自己在学生时期就曾探索过的马列维奇的建构(第18页)的空间性质。在为这次展览所做的设计中,有两座大型构造别具特色,其中一座是泰特林之塔(Tatlin Tower),另一座便是马列维奇的构建。这两座设计均以独特的方式致敬弗兰克·劳埃德·赖特的螺旋式结构,又均被空间所扭曲。马列维奇的建构第一次变得宜居了:想进入上层的展馆,游客们需从这个建构中穿过。

这项设计让展品之间的对抗得以凸显出来。比如,泰特林之塔与马列维奇的建构之间建立了一种对抗——马列维奇的《红场》与弗拉基米尔·塔特林(Vladimir Tatlin)的《转角离隙》的对抗。展厅中展览着来自于"展示0.10"的作品,在这片展览空间中,马列维奇的建构的一部分被抬离了地面;而在黑屋子里则展览着来自1921年的"5×5=25"展览的展品,那些摆在有机玻璃展台上的绘图看起来似乎是悬浮于地面之上。

伟大的乌托邦(THE GREAT UTOPIA)
美国,纽约,1992年

马德里美景（VISION FOR MADRID）

西班牙，马德里，1992年

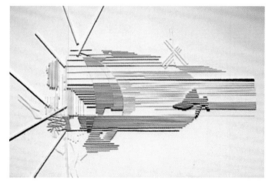

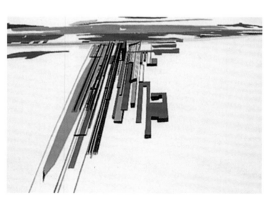

  自古以来，马德里的发展就像一层接着一层张开的贝壳：先是圆形的中世纪城市，然后是19世纪的棋盘式布局，进入20世纪之后，一条新月形的高速公路如今带动着城市呈线性发展。西部的曼萨纳雷斯河将城市的发展拦截，现今它主要是往东发展。郊区的住宅建筑如蘑菇一般一个接一个地冒出，它们越过了M30号高速公路，正欲气势汹汹地吞没附近的村庄。

  城市正在失去自身应有的形状，我们的目标是阻止它，我们要疏导和阻止这种无法无天的发展和蔓延。为了达到这一目的，我们设计了四处独特的区域用于城市的发展和更新。在南部，从前沿着铁道而建的工业建筑将被改造成生气勃勃的公园和休闲景区，新的商业区将沿着带状廊道集中发展，一直蔓延到机场所在的位置。而作为从北向南的轴线，埃尔卡斯蒂利亚大道将变得越来越繁华，新的建筑将穿插在鳞次栉比的建筑之间，而露天的小块空地将成为公共活动的空间。最终，郊区余留的空隙将得以保留下来。

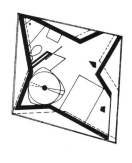 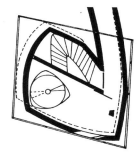 

(从左至右)第一层(星形),第二层(螺旋形),第三层(十字形)

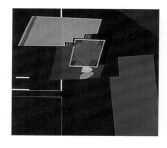 

这座扩建而成的艺术酒店的原型非常有趣:一座半木质的农舍和一间马厩。员工们想将它改造成极具挑战性的新建筑构造,使其成为雕塑般的存在。作为"凝聚体",这项设计的中心是一片椭圆形的空间,它协调着旧的建筑和新的建筑。这片空间部分延伸至地下,它可以被用作展览或表演。在轴心点之上的塔楼里有三层房间,它们建造在十字构造和星形构造之上,每一间房间之中都是完全自成体系的环境,其内部还装有嵌入式的家具和配置。作为第三个主体,螺旋结构被另外两处空间所共享,它们通过螺旋状结构与主楼相连。在这样一个极具当地特色的村落里,这套建筑营造出对比强烈的效果,它似乎意在鼓励人们思辨和创新。

比利·斯特劳斯艺术酒店(ARTHOTEL BILLIE STRAUSS)
德国,莱朋,1992年

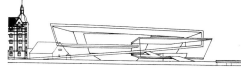

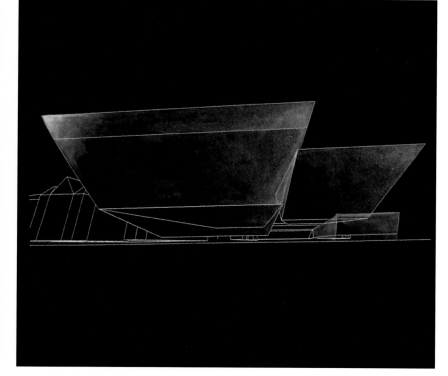

阳光、地面与水流深深地切入这一实体的建筑之中,它们完整地填满了施工场地。这一大型建筑不仅具备紧凑感,而且还具有开放性。这些锐利的切线朝向天空,彻底地解放了大厅的张力。一道弧形的斜切线将公共广场一分为二,将人行道缓缓引入了建筑之中。如同雕塑一样的建筑极具表现性,由不同色泽的花岗岩构成,一座座建筑之间是被压缩的空间。这些建筑构造由就地浇筑的加固混凝土构成,形成了不同形状的平板。

音乐厅(CONCERT HALL)
丹麦,哥本哈根,1992—1993年
与帕特里克·舒马赫合作

为了将这一前工业区与城市相连，我们应用了三项典型的构造手段——梯形结构、楔状结构以及螺旋状结构——来定义这一多功能场地，并将其与周围多样的环境融合在一起。这些构造是巨大的、模糊的实体，被按比重置于建筑物与地形之间。它们一起构建了连贯的区域，这里遍布着高密度的文化、休闲、住宅以及商业场所。旧建筑也被赋予了多样的功能以及新颖的结构。

梯形的区域覆盖了整个海港盆地，位于码头周围的两座建筑中建有航船设备以及河船入港检验中心，其北部建有会议中心。在梯形区域与楔形区域之间拔地而起的是倾斜的办公室塔楼。楔形区域从莱茵河岸一直延伸至乌别琳，将河边地带与雪华铃住宅区（Severins residential quarter）连成一片。这里的房屋被建造成了立于高处的水平排房，它们连成了长长的一排，看起来就像是从前的仓库。它们似乎是被有意地抬高了，以便于人们能够一览无余地观望河上的风景。螺旋状结构将罗默公园（Römerpark）与码头周围连成了一片，河边公路的一段便从其周围绕过。在这片建筑场地上，新的文化中心如同零散的宝石折射着摇曳的水波，而河流则常年从这里流过，随着交替的四季而不断地变化着。

场地规划图

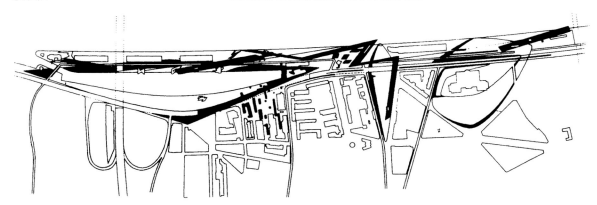

科隆住宅小区（RHEINAUHAFEN REDEVELOPMENT）
德国，科隆，1992年

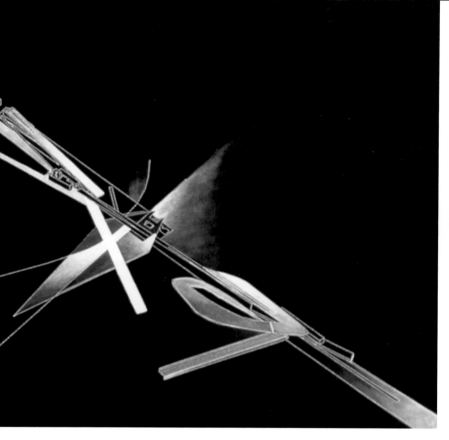
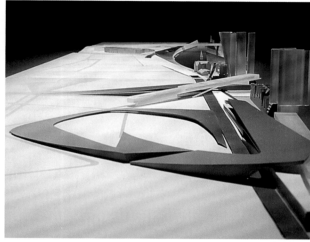

在这一建筑场地上,人类文明的痕迹不仅与风景融为一体,而且似乎与环境也是难舍难分的,从而为这里增添了一处迷人的特色。按照这种思路,我们打算将山边主题公园的建筑——内含一座地质中心、一座户外剧场、一座观景台以及一座博物馆——塑造成另一处延伸的人造景观。这些建筑是新文化最初的零散展现,随着时间的推移,它们将覆盖周围的港湾。这些零散的建筑预示着一种截然相反的建筑风格。

地质中心仿佛就栖息在山脉自身的历史之中,它如同一把利刃一般切入地层之中,向我们展现着基岩。基底如同断层面一般倾斜着与彼此交织,一个切面切入山脉之中,其他切面也顺势紧随其后。户外剧院如同一座希腊的圆形剧场。它就像一件"浑然天成的艺术品",立于菱形玻璃展台之上,而玻璃展台则是顺应其下面的地势建成的。一座凸起的建筑从这一凹陷的空间中拔地而起,它悬挂在斜坡之上,随着时间的流逝在建筑场地的展示台上结晶成形。这座博物馆合成了其他三项设计的理念。巨大的平板如同露出地面的矿脉一般,从地面层向上竖起、伸入空中。

卡农顿(CARNUNTUM)
奥地利,维也纳,1993年

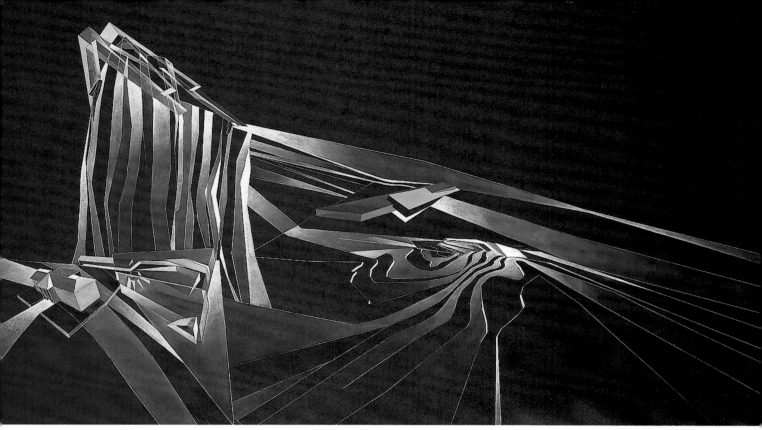
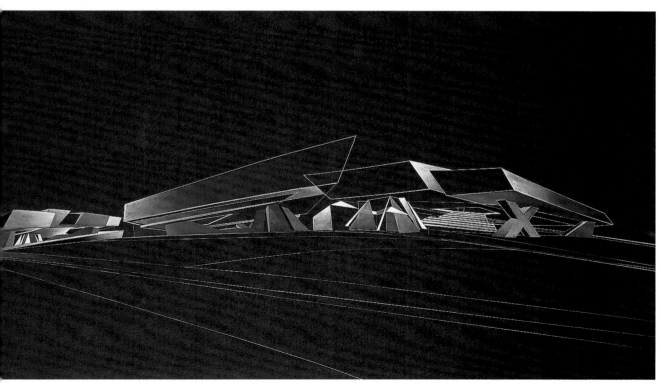

卡迪夫湾歌剧院（CARDIFF BAY OPERA HOUSE）

英国，卡迪夫，1994—1996年

这项设计意在展示城市设计中两种通常相互排斥的方式：无限的雄伟与有限的空间。该项目是一大片建筑物的一部分，它们共同构成了椭圆形盆地广场。但这座建筑构造本身就是海滨地区的一大标志。典型的街坊建筑体现着其自身的矛盾性，它向外塑造出更大的公共城市空间，向内笼罩下与世隔绝的内部空间，而这两类空间相互碰撞所形成的真空则消解了建筑自身的矛盾性。这一效果是通过三种补充策略实现的：一是抬高边界；二是在拐角处指向码头外端的边界处打开一个缺口，以展现厅堂的构造，使其作为主要的实体立体形立于所圈定的场地之内；三是通过一道缓缓伸入建筑场地的斜坡拓展广场的范围，从而使公共空间得以延伸，以便在主门厅区域铺设新的地平面。

于是，这项设计构造出了高于地面的广场，它不仅仅可用于露天表演，而且还能使观众在回头时看到内港以及海湾的景色。这座建筑的建造理念基于以建筑的形式表现服务空间与被服务空间之间的等级。观众席以及其他公开、半公开的表演、彩排空间像宝石一样从一排整齐排列的支架上脱颖而出。这一排支架环绕场地一周，像一条内翻的项链一般将整个场地包围。其上的"宝石"在彼此之间围起集中的公共空间，人们可以从这里直达剧场中心，而人们所需要的服务则由其身后的剧院边界处提供。中心处是露天的中庭，其底层之下是门厅区域。凸起的底层上下均有观众席和主排练室。两条轴线从两个主入口处伸出，它们交叉穿过剧院中的空心，延伸着一直切入整个平面。这两个入口分别位于椭圆形盆地广场以及码头街广场之上。

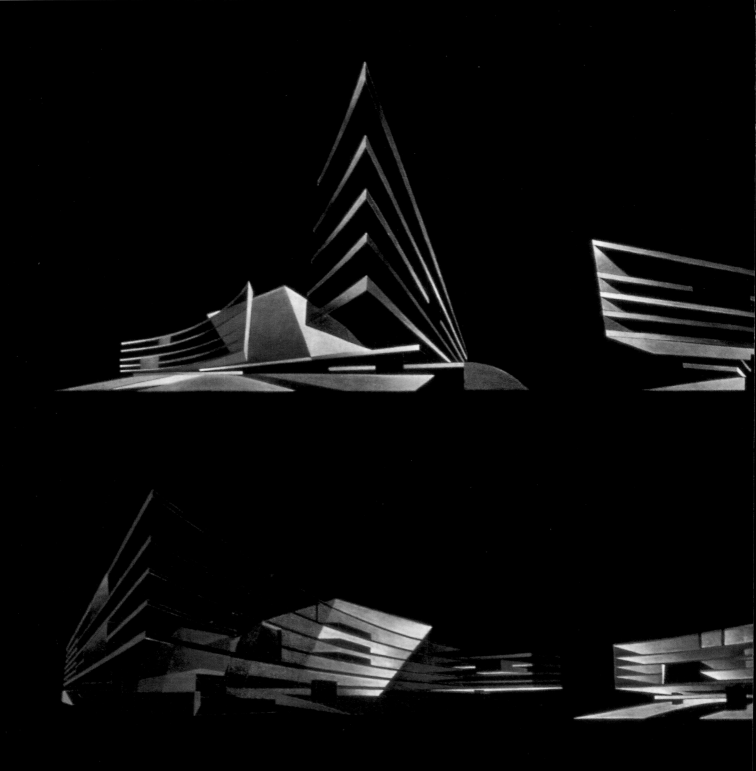

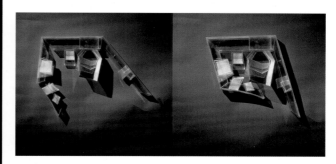

卡迪夫湾歌剧院

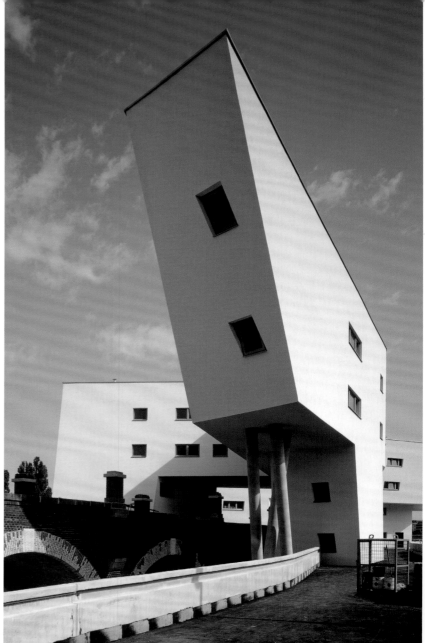

斯皮特劳高架桥（SPITTELAU VIADUCTS）

奥地利，维也纳，1994—2005年

　　作为维也纳市振兴维纳地带（Wiener Gürtel）计划的组成部分，环状的建筑结构切入城市的肌理之中。这一拥堵的区域布满了相互遮盖重叠的基础设施，其中包括维也纳最为拥堵的街道之一，多瑙河运河以及穿梭于银行间的自行车道，还有威尼斯铁路系统的历史残留痕迹，尤其是还有一座由奥托·瓦格纳（Otto Wagner）设计的高架桥。一排排的公寓、办公室以及艺术家工作室如丝带一般生动地围绕着、俯瞰着高架桥那弯弯的圆拱，营造出一种与众不同的室内和室外的空间关系。屋顶则是私人活动场所，为繁忙的运河之畔增添了生机盎然的一笔。

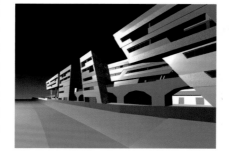

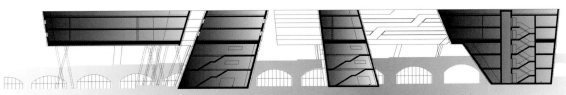

蓝图展览馆，1995年国际建筑博览会（BLUEPRINT PAVILION, INTERBUILD 95）

英国，伯明翰，1995年

与鲍尔·布里斯琳（PAUL BRISLIN）和姚伍迪（WOODY YAO）合作

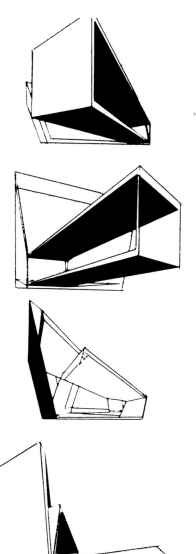

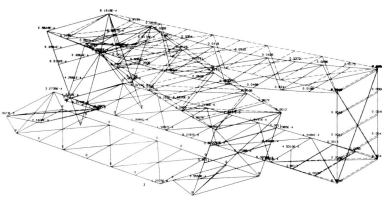

这座展览馆的形状被设计成了不断延展的空间，它将展品连贯在一起，使参观者们能够环绕着展馆参观。展馆的结构由一面不断延展的平板构成，它向外延伸着并反折过来形成了两面相互交错的厚板，于是厚板的内侧变成了展览的空间。这块平板的底面为金属厚板，厚板的外层包裹着镀层（铝板或者工业折叠板），厚板的内层也包裹着镀层（中密度纤维板、工业地板或者其他装饰材料）。以这样一种方式，可以将装饰品畅通无阻地从地板一直铺设到周围的墙壁上。灯光被置于平板内部深处或者是遍布于平板上方。总之，这条如同莫比乌斯带（Möbius strip）的平板塑造了完整而彻底的展览空间。每一位参展者在这一"物体"中都被分配了特殊的展区，而展品则流畅地从一个空间排列至另一个空间。

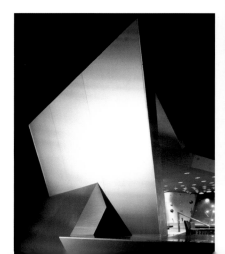

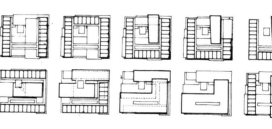

我们在第八大道的第四十二街设计了一座酒店综合大楼。我们意在设立一处城市缩影,以彰显一座国际都市的精致与魔力。所设计的综合大楼由两座三层的商业裙房和两座酒店构成,其北部建筑为45层,其南部建筑为22层。循环系统、动线标志、照明设计、相关娱乐配套设施,以及零售业务使这座建筑活跃起来。酒店则如同一条被竖起来的街区——这座高于一般高层建筑的酒店总共有950间房。第二层的建筑元素伸入主楼中心的空地之中。在酒店塔楼与商业裙房相连接的地方,垂直街区式的塔楼上则延伸出一面水平的建筑面——其上分布着零售商店、酒店以及公共设施——它们与城市相接,一直延伸至其下的地下广场。

第四十二街酒店(42ND STREET HOTEL)
美国, 纽约, 1995年

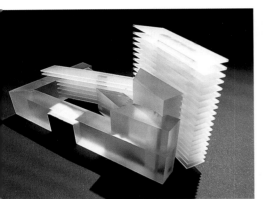

按照要求,我们应沿着莱比锡大街设计一座办公楼。处于主要街道的交错位置,在柏林市的米特区最繁华的路段之一,与这座城市中的维多利亚城市区域项目(第40页至第41页)所展现给我们的景象不无相似之处。在这座建筑中,不仅设有大型金融机构的总部,还设有私人办公空间。我们的设计意在调和19世纪的建筑与战后的建筑,在推倒柏林墙之前,战后那些毫无特色的建筑结构曾占据了天际线。我们将两个"L"造型的构造相互盘绕、交织在一起,而这座建筑则由三层主平板构成。它们仿佛处于车水马龙的十字路口上,这些庞大的结构倒映着周围的环境,仿佛使经过的车流更加川流不息。

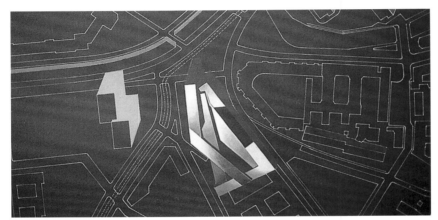

场地设计图

斯皮特尔集市(SPITTALMARKT)
德国,柏林,1995年
与帕特里克·舒马赫合作

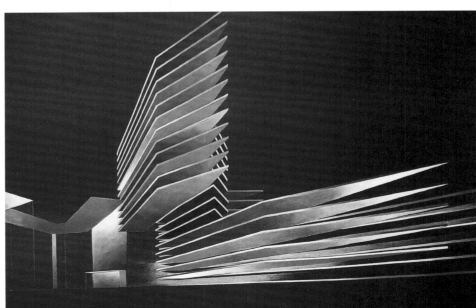

南肯辛顿一所学校的所有者让我们为他建一座门房以及守门人的住处，它们应位于学校的入口处，入口则通向一座小型建筑物及其后院。我们决定旋转传统的垂直门，使它的造型为一块浮动的水平板，从而与房子融为一体。学生们可以穿过数根森林般的圆柱体系统，自由地出入这座建筑，就像平常穿过一道门一样。随后，被抬高的水平板便会碎成一片片谜一般的碎板，而每一片碎板都由其结构系统中的圆柱、翼片以及墩座所支撑，从而将底层变换为尽情施展结构元素的舞台。除了门房及其家人的居住区，这座建筑中还包含四间教室。

戴高乐法语学校（LYCÉE FRANÇAIS CHARLES DE GAULLE）
英国，伦敦，1995年

因为伦敦的金融区有着独特的历史环境，所以想要在拥挤的建筑场地上修建一座办公大厦，难免会遇到各种各样的规划以及建筑限制。直接影响到我们的一项限制便是底层必须包含一处离街的公共区域，它将发挥室内公园的作用。我们建造了一座建筑物，使它的结构环绕建筑场地一周，从而塑造出一间"露天房间"，如此我们便符合了这项限制的要求。在施工场地所允许的范围内，这一"蛇形"建筑在室内空间与室外空间之间，在私人办公室与公共广场之间达成了一种平衡；与此同时，它还与这座城市的传统建筑形成了动态的互动。

潘克拉斯小巷（PANCRAS LANE）
英国，伦敦，1996年

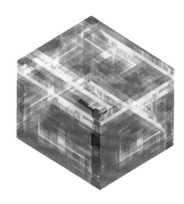
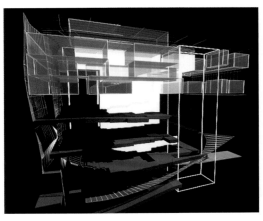
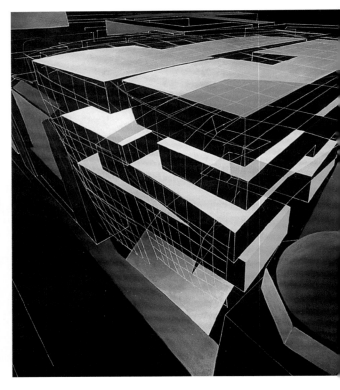

锅炉房扩建（BOILERHOUSE EXTENSION）

英国，伦敦，1996年

维多利亚和埃伯特博物馆（The Victoria & Albert Museum）如同一个大杂烩，它是由各时期的建筑物拼缀而成的。在过去的150年中，它曾以这样混杂的模式发展。在经过一段时间的向外扩张之后，人们决定塑造博物馆的内部，有效利用其剩余的内部闲余空间。于是，我们为这座前锅炉房设计了方案，这座博物馆成了我们的代言人，代表着经我们之手所改造的建筑物。无论是塑造显示面板，还是构建展览空间，我们均以像素的元素为媒介设置结构。最上面三层为相互交织的建筑体量，其中建有管理设施、教育与活动中心以及机械设备间，它们与这座博物馆已建的侧厅相连。在这些实体建筑之间，屋顶与立面上均开有空隙，以便阳光照射进来；原有的立面与新建筑之间的区域上也嵌入了空隙，从而使我们依旧能看见阿斯顿·韦伯（Aston Webb）所设计的立面。建筑的立面由两种铺板构成，它们各自具备独特的功能。它们时而相互交织，时而融合在一起，构造出地板、墙壁和窗户。起伏的外层铺板是由玻璃以及金属平板所构成的一层防雨板，而内层铺板上则设有百叶窗，从而为展览遮挡阳光、控制光线。

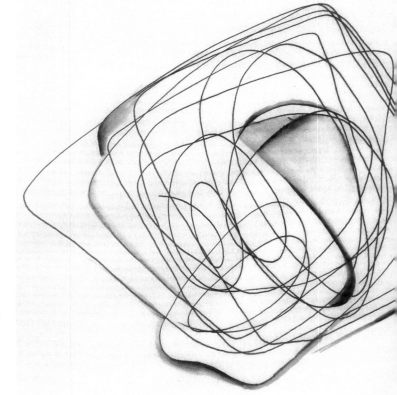

在解读建筑时，我们不能将人文科学中现实与虚幻的混乱缩减为完美的柏拉图式的形式，没有什么是被优先考虑的。如此，数面墙壁构成了位于维也纳艺术馆的展览空间，营造出多角度的模糊效果。这些墙壁从入口处开始向着各个方向延展，穿过了整间盒子一样的展览室，而展览室几乎容纳不下它们。这里没有预先规定好的路径，你随时都会与惊喜不期而遇。

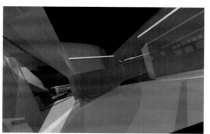

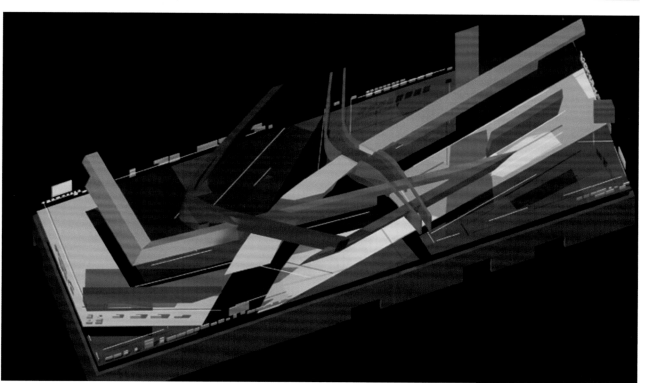

许愿机器：世界创造（WISH MACHINE: WORLD INVENTION）
奥地利，维也纳，艺术馆，1996年

大师专属：威尼斯国际艺术双年展
(MASTER'S SECTION, VENICE BIENNALE)

意大利, 威尼斯, 1996年

这个画廊展区在格拉西宫美术馆 (Palazzo Grassi) 里, 由一间椭圆形的房间和一座小阳台构成。因为这间房的四面均与主要的环路相连接, 所以我们打算突出房间的容纳空间。我们并没有阻断游客流, 而是通过将墙壁提升至高于地面2米的高度, 释放了墙壁展示区的空间。绘画、素描、模型以及浮雕被展示在墙上, 它们以这样的方式为我们创造出一幅"超级景象"。

宜居桥 (HABITABLE BRIDGE)

英国, 伦敦, 1996年

这座桥被塑造成了一座躺下的摩天大楼, 它极有可能成为这座城市中的主要景致之一。这座桥位于这座城市的文化混杂区域, 它将一系列的活动以及功能——零售、文化和娱乐融合在一起, 融入这座建筑的结构之中。我们为伦敦的这座桥设计了方案, 它的独特之处在于附于桥上的一片片建筑板, 它们一起倒映着河畔熙攘的城市生活。建筑板在与彼此相接的地方断折开来, 从而形成一系列的空间和路径, 它们均指向河岸所在的地方。桥梁的走势并非一成不变, 转折处更是突出了建筑板断折的效果, 从而使得人们能够沿着东西轴线看风景, 使人们畅通无阻地看到从西南方的里士满一直到东方的圣保罗的景色。这座桥梁是按照垂直模式构建的, 下层分布着畅通无阻的公共通行"街道"——商业区与文化区的混合体, 而桥梁上部的空间中则分布着阁楼一般的私人空间。空间与路径共同构成了流动的整体, 将河流的风韵发挥到极致。

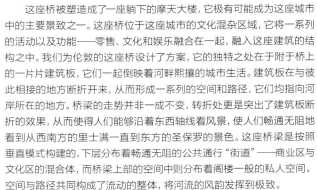

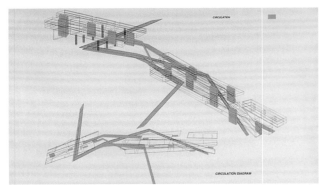

凤凰剧院（LA FENICE）
意大利，威尼斯，1996年

1996年1月，一场大火吞噬了意大利一座最负盛名的歌剧院，英国报纸《每日电讯报》委任我们设计补救方案。威尼斯是一座以塔楼著称的城市，其建筑的屋顶上布满了烟筒和尖顶；依照这种风格，我们预想修建一座高耸的歌剧院，为城市屋顶的固有特色再加深一笔色彩。因为威尼斯这座城市本身就是一座剧院，我们决定以逆常规而行的方式设计这座凤凰剧院。我们决定向外界展示表演，塑造了露天的舞台以及观众席，它们面向广场以及运河。运河的河畔区域将被改造成观众席，而位于其后的歌剧院正面将被装上投射屏，正如我们为卡迪夫湾歌剧院（第70页至第73页）所做的设计那样，当你立于不同高度的门厅之内，便可以看到从观众席到下面广场上的景色。人们沿着墙壁攀上露天的楼厅，以便更好地观看底层的表演，他们有时候也会直接攀爬到观众席之间的包厢内。

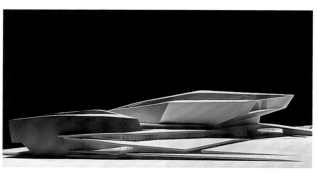
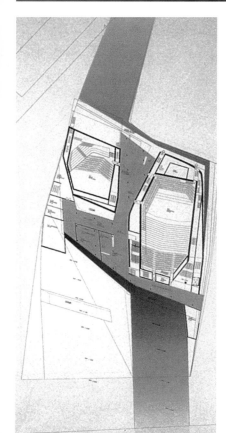

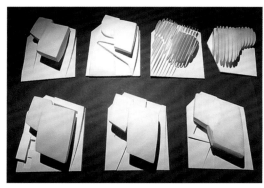

面向旧城的陡峭山体为我们提供了创意，使我们形成了这项设计的概念：风景。一系列层叠的、梯状的、倾斜的地面与屋顶激发了我们的灵感，让我们决定塑造一处起伏的建筑场地。一座巨大的厅堂以及一座演艺厅在这一"风景"之上拔地而起。一道缓缓起伏的斜坡通往前部的大厅、包厢和休息室，它引导着游客们一直走入建筑物的内部。这些斜坡和坡道构成了绵延不绝、起伏不断的风景，庭院则出现在恰到好处的位置，为底层的表演活动提供了光线。内部的装修将风景进一步延续，以波澜起伏的地势定义着循环的空间。

爱乐厅（PHILHARMONIC HALL）
卢森堡，1997年
与帕特里克·舒马赫合作

1999年园艺展（LANDESGARTENSCHAU 1999）

德国，莱茵河畔魏尔，1996—1999年
与帕特里克·舒马赫和迈尔·巴赫尔（MAYER BÄHRLE）合作

　　这座园艺展的展厅是一系列设计中的一部分，我们试图通过钻研自然景观的构成，在这些设计中塑造出流动的空间性，诸如河谷、山峦、森林、沙漠、峡谷、冰川和海洋。与都市以及建筑空间不同，这些景观空间最重要的特点是能够巧妙地划分区域并流畅地过渡风景——从而使空间以虽错综复杂却差别细微的方式——呈现。在建筑物自然地铺设、分离和接合之处，景观也随之打开、呈现和暗示。

　　建筑空间聚拢之处恰恰彰显着风景最为自由的侧面。这座建筑并没有被隐没在风景之内，而是脱颖而出，它从交织的道路之间一点点地浮现出来，让游客自己去分辨哪里是风景、哪里是建筑。公共道路绕过建筑物、绕过切入地下的门廊，使人难以分清哪里是"地面"。柱子参差不齐地排列在下层的地面之上，与上层中天花板之下的柱子遥相呼应，建筑的表面呈现着接连不断的几何结构。于是，压缩的空间里布满了重叠的形状，它们给人以视觉上的飘荡感。如同两大阵营一般，节奏和质地之间在相互对抗。

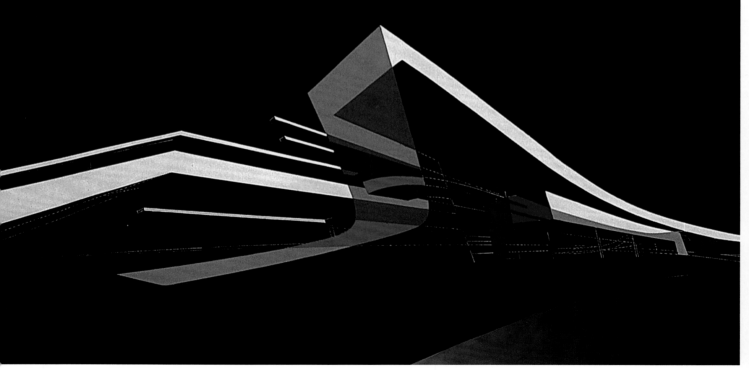

1999年园艺展

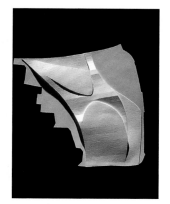

伊斯兰艺术博物馆 (MUSEUM OF ISLAMIC ARTS)

卡塔尔,多哈,1997年
与帕特里克·舒马赫和姚伍迪合作

在中东,不曾有过鲜明的19世纪风格的博物馆。所以,我们为这一类型的建筑开了先河。我们立足于穆斯林文化中对于重复形状的偏爱,以及时而出现的变化调和建筑的形态。从整体上来看,这座建筑就是一个容器,其中容纳着依次登场的"展览品"。一个概念回荡在展览空间中:在这座呈阶梯状分布的大展馆里,水平与倾斜的木板搭建出了展览的空间;从钱币到手稿、玻璃仪器、地毯,这里展览着琳琅满目的艺术品。在这座建筑的建造理念中,风景发挥了至关重要的作用,尤在于它试图将建筑元素与环境天衣无缝地融合在一起。屋顶具备定义全局的功能,它一方面将这座建筑表达为无限延展、千变万化的空间,另一方面又调和着风景、天空与展厅之间的关系。伸入建筑内部的庭院提供了自然的光线,这符合阿尔菲纳(Al-fina)的传统,与伊斯兰建筑以及设计两相呼应。

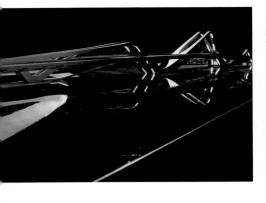

谢赫扎耶德大桥（SHEIKH ZAYED BRIDGE）

阿拉伯联合酋长国，阿布扎比，1997—2010年

阿拉伯联合酋长国是一个交通高度发达的国家。他们打算环绕着波斯湾的南岸建造一条路线，从而将七个酋长国贯通在一起。1967年，他们搭建了一座钢铁拱桥，将阿布扎比岛与大陆连接在一起；随后，第二座桥梁建于20世纪70年代，它贯通了岛屿南岸的下游。通关口岸靠近第一座桥梁，对于整个道路系统来说，它有着至关重要的作用。这座桥梁立于四通八达的位置上，它本身就具备潜力成为一处目的地，从而促进阿布扎比的城市发展。一组铁索的一端被固定到岸边，然后被提起来贯穿整个海峡。一个"正弦波"状的造型构成了桥梁的结构轮廓。大陆这一端是整座桥梁的起点，它从地面向上拱起与道路相连。路面悬于脊骨梁结构的两边，在两侧公路中间巨大的混凝土基墩上，钢铁的拱形以不规则的样式升起，这样人们便可以分清哪里是陆地、哪里是航道。主梁在一岸一分为二，沿着中空的位置向前伸展，在路面之下一直延伸，直至桥梁另一端的路面之上。桥梁的主拱高出水平面60米，路面的最高处高出平均水位20米。

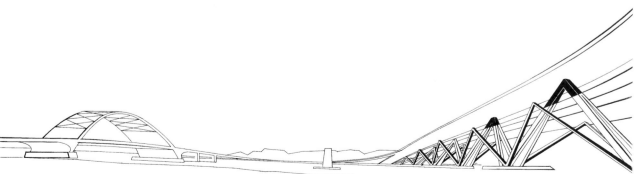

伊利诺伊理工大学校园中心 (CAMPUS CENTER, ILLINOIS INSTITUTE OF TECHNOLOGY)

美国,伊利诺伊州,芝加哥 1998年
与帕特里克·舒马赫合作

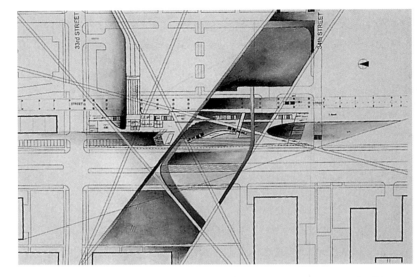

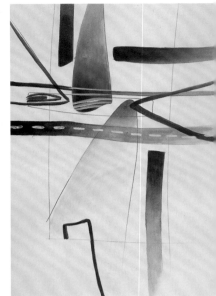

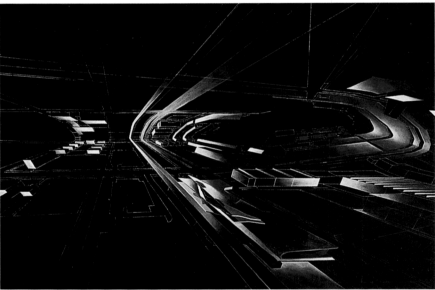

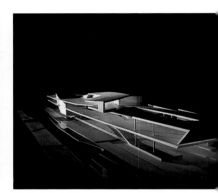

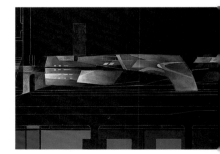

　　在这项设计中,我们在密斯·凡·德·罗的校园中建造了一座学生中心。这向我们提供了一个机会,让我们能够致敬大学内部社团的多功能模式和芝加哥自身的结构。我们决定构建流线型的组织架构,以模糊工作区域和休闲区域的界限,将横向布局的总体规划设计立体化,从而确保校园的元素能够成为紧凑而多层次的立体构造。这座建筑物的楼地面与弯曲的斜坡相交的地方,是一座双层的前厅,它将顾客引向礼堂、咖啡厅和零售空间。第二层与第一层相互错开一部分,被楼梯的斜坡一分为二,留下空隙使人们能够看到下面的风景。会议室是由滚动面板构成的矩阵,根据学生们的需要,这些滚动面板可以自由地开合。所有这些空间中都没有布置清晰的界限,以便鼓励学生们进行跨领域的活动;模块化系统的台面更是提高了空间的灵活性,它们能够组合出令人出乎意料的效果。

洛伊斯和理查德·罗森塔尔当代艺术中心 (LOIS & RICHARD ROSENTHAL CENTER FOR CONTEMPORARY ART)

美国，俄亥俄州，辛辛那提，1997—2003年

　　这座博物馆建于1939年，它跻身于美国第一批致力于表现当代视觉艺术的建筑之中。而建造这座新建筑的目的是为了吸引行人和创造动态的公共空间。位于第六街和胡桃街的拐角处，这座建筑的底面慢慢向上卷起，延伸至建筑的主体，成了建筑的后墙。作为一张"都市地毯"，它鼓励着行人走入入口、进入大厅之中，引导着行人走过整个长廊，走上悬浮的夹层坡道；而整个长廊在白天就是一座开放的日光"景观"或是人造公园。

　　与都市地毯那抛光的起伏表面形成对比的是展廊，它们悬浮于大厅之上，仿佛是一整块大理石镂刻出来的。展览空间有大有小、形态各异，适用于多种规模以及材料的当代艺术。当你穿过建筑后部的一条狭窄裂缝，走上弯曲的楼梯坡道，当你走过循环往复的通道进入画廊的时候，你所看到的景象将会是出乎意料的。这些变幻多端的画廊交织在一起，如同立体的七巧板一般，在建筑中构建出虚与实。

　　由于建筑物位于街角，我们便为其设计了两面各不相同、相互补充的立面。朝向南方的立面位于第六街上，它的表面呈半透明状，由此行人们能够看清建筑内部的活动。朝向东方的立面则位于胡桃街上，它的表面呈浮雕状，以顺应画廊内部凹陷的构造。

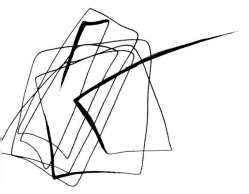

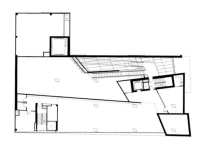
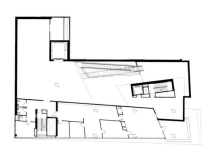

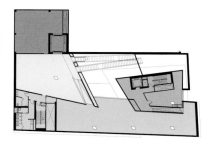

思想地带(THE MIND ZONE)
英国,伦敦,1998—2000年

位于千禧穹顶的内部,思想地带是14个独立展览空间之一。为了完成这一独特的任务,我们设计了管理与建筑计划。这项设计以相互重叠的三层结构表达了复杂的思想主题。三层结构依次打开,塑造出不断延伸的表面,地板、墙壁和拱腹一一展现在你的眼前,你可以在空间中流畅地穿行。从地板到墙壁均由轻型透明木板铺设而成,表面覆盖着玻璃纤维,内部为铝制蜂窝状结构。同样,钢基上也覆盖着一层透明的材料,以便营造出转瞬即逝的效果,这样的创意特别适合一场为期仅仅一年的展览。

霍洛威公路桥 (HOLLOWAY ROAD BRIDGE)
英国，伦敦，1998年

当时的北伦敦大学的校园散布于一片城市地带，它覆盖了霍洛威公路、铁路和地铁。在这座城市的交通网中，建造着地面之上的第二层交通通道，而它们构成了这座大学自身内部循环模式的一部分。我们所设计的这座桥梁则聚拢并拓展了这一内部交通网，将其自身与现存的建筑中心及路径贯连起来。桥梁慢慢延展，它不断适应着新的地形，满足着学校的多种需求。空中大堂位于桥梁的着陆点上，这里分布着咖啡厅、图书馆和研讨室。这座桥梁的结构是由钢桁架构成，架子的支撑以及覆盖模式给人以交织错落的感觉，从而为桥梁内部营造了交错的光影。

宠物店男孩世界巡演舞台设计 (PET SHOP BOYS WORLD TOUR)
1999年

建筑与音乐表演并不总能融合在一起，因为它们适用于不同类型的规则。然而，为英国流行音乐组合"宠物店男孩"所设计的舞台，却是一项极具挑战性以及启发性的作品。作为两项规则杂糅的产物，它既符合建筑的规则，又符合音乐表演的规则。在设计这座舞台时，我们并没有有序地安排空间，而是以一块铺开的白幕布来为这活跃的音乐会添彩。一面不断延伸的立面被弯折、分离，构成了背景、架构与台面。这一立面的其他部分则由附着于其上的移动元素构成，它们如同舞台工具一般，在灯火通明的立体风景中，投射着灯光，制造着音响效果。

停车场和奥埃南诺德总站
(CAR PARK AND TERMINUS HOENHEIM-NORD)
法国,斯特拉斯堡,1998—2001年

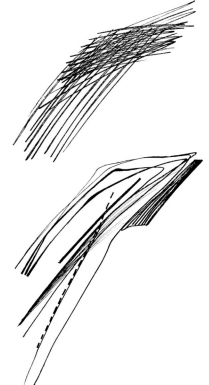

  日益繁忙的生活加重了市中心的拥堵与污染,于是城市中新设了一项电车服务,以鼓励司机们将车留在专门为他们所设计的停车场里,转而改乘电车回家。这座拥有700个停车位的停车场位于城市线路的最北端。它的建造理念是将相互重叠的"场地"以及线路编织在一起,以构成一个多样的整体。每一种移动方式——汽车、电车、自行车和步行——都被分配有各自的轨道和路线,车站、风景与环境中的材料和空间不断变化着,似乎以此呈现着不同交通模式的转换。

  空间的安排提升了路线的三维立体感:地板上的光线、家具装饰线或者天花板上的条形照明灯光线,这些线条彼此交织,绵延不绝。按照所设计的规划,所有这些"线条"相互交织,构成了一个难分彼此的整体。在停车场里的汽车如同不断变换的元素,人们在黑色的柏油面上画下白色的线条,这样每一处"磁力场"都是已圈定的停车位。停车场位于场地最低洼的地方,南北朝向依次排开,然后随着场地界限的弯曲而轻轻地排成环状,每一处停车位上均设有照明设备,保持着统一协调的高度,以适用于倾斜的地面。

  作为配套设施,电车站和停车场使地面、灯光和空间之间形成了一种共鸣。露天风景空间与公共内部空间之间的界限模糊,通过把握二者变换的瞬间,一种新的"人造自然"诞生了。它不仅模糊了自然环境与人文环境之间的界限,而且改善了斯特拉斯堡乘客的日常生活。

停车场和奥埃南诺德总站

国立二十一世纪艺术博物馆
(MAXXI: NATIONAL MUSEUM OF XXI CENTURY ARTS)

意大利,罗马,1998—2009年
与帕特里克·舒马赫合作

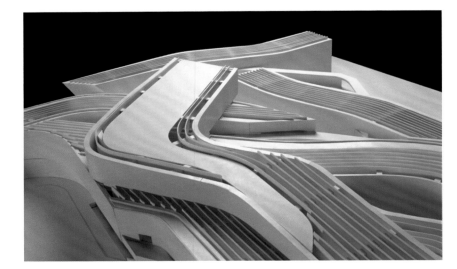

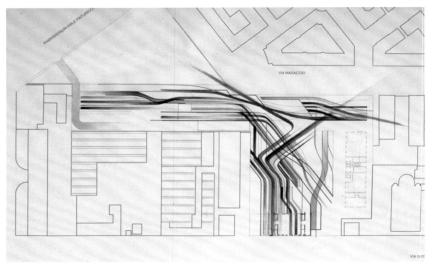

  在设计这座博物馆时,我们并没有遵循类型学的模仿原则,而是顺应了四周并不高的城市肌理,从而与临近较高的水平建筑形成了鲜明的对比。于是,这座博物馆成为了一座"城市嫁接物",如同此处的第二幅面孔。这座建筑将台伯河与维亚·圭多·雷尼街相连,其本身便蕴含着流动的样式,无论是显性与隐性,抑或是内在与外在。它将循环的概念应用到城市的环境之中,这座建筑与城市共享着公共的空间,覆盖了带状的道路以及露天的空间。除了循环的关系之外,建筑元素在形状上还顺应了此处城市的棋盘式布局。

  这项设计注重塑造准城市化的场地,而并非在定向草图的基础上组织和操纵建筑物;它分散了稠密的布局,并不专注于突出博物馆的造型,而是反映了博物馆的整体特点:一处透气的、沉浸感强的空间。垂直与倾斜的循环元素被应用于交汇、扰乱和躁动的空间之中。对于理解建筑与内部艺术品之间的关系,从物体到场地的变动有着至关重要的作用。这项设计鼓励人们颠覆以"物体"为本源的展览空间,从而以具体的形式表现"流动的"概念。于是,"流动的概念"不仅成为了建筑的主题,也成了依据常规穿行在博物馆中的方式,这种方式将参观者带离"物体",使之融入各种交汇的场地之中,迎面而来的则是无法逃避的变化。

皇室珍藏博物馆（MUSEUM FOR THE ROYAL COLLECTION）

西班牙，马德里，1999年
与帕特里克·舒马赫合作

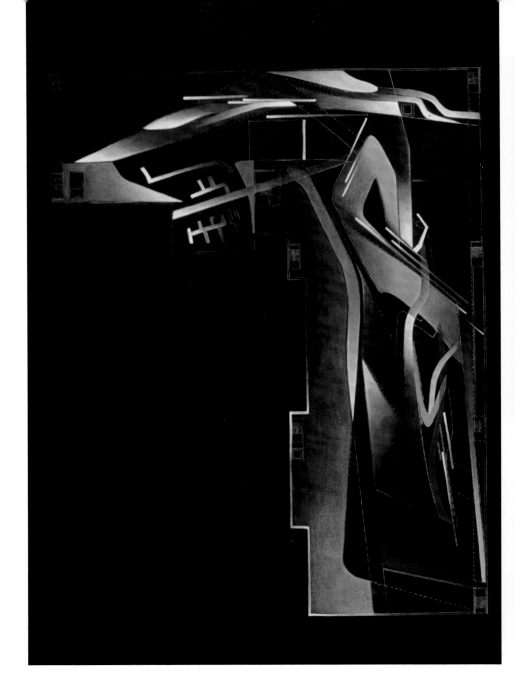

在设计这座位于马德里的新博物馆时，侵蚀形态为我们提供了灵感来源，因此所有的构造都非实体的形状，而是凹陷的围合空洞。设计地点位于皇宫和大教堂之间，这座建筑的大部分体积位于地下，仅仅在广场上建筑的缝隙间，能瞥见它的几丝剪影及其表面的一丝侧影。一面宽大的三维立体墙撑起了其内部的结构，沿着两个空间环绕迂回着穿过。建筑空间中的可用容积并没有被分层，而是被完整地保留在了整座博物馆中。可以看见两条蜿蜒的道路相互交织，展现出严密的线性顺序。空间中的垂直动态是通过梯形结构实现的，从而展现出宽广的表面。这些露台就像是流畅的斜坡，沿着中空的结构，使空间内的空气得以循环流通。

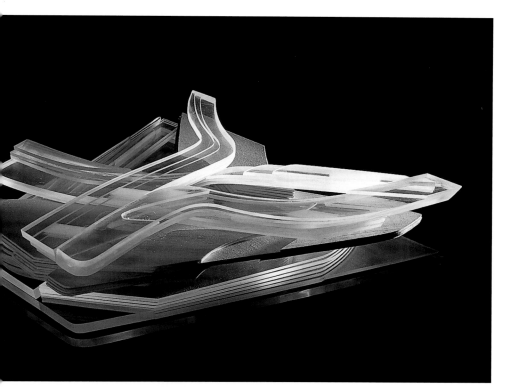

我们最初的目的是要为这座博物馆设计出一座一眼就能认出的建筑。它应具备独特的外形,而灵活性则是我们所要达到的主要目标之一。展馆之间由辨识度很高的路径相连接,并且很好地接合了各展馆内的空间。沿路穿行在展馆之中,你会拥有不同寻常的体验。一层层的建筑层交错着倾斜,露出一道道的缝隙,从而将顶部的自然光线发挥到极致。公共入口设在建筑东面,将建筑清晰地分为公共区域和私人区域,而西面则被用于运输。

雷纳·索菲亚博物馆扩建(REINA SOFÍA MUSEUM EXTENSION)

西班牙,马德里,1999年
与帕特里克·舒马赫合作

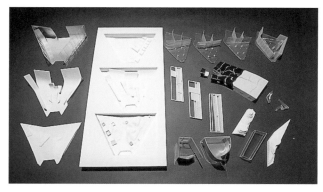

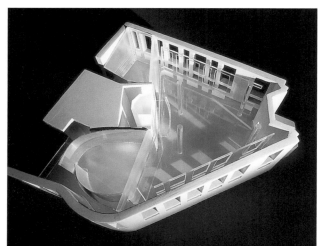

罗斯柴尔德银行在伦敦设立了新的总部,我们要为其门厅设计内部装饰,一并设计接待处以及展示柜。我们为其设计出了三种简明扼要的方案:第一种方案是将入口处设计成玻璃房中房,它与周围的环境截然分开;第二种方案是将空间塑造成绵延不绝的表面,墙壁与地板之间并没有明显的界限;第三种方案是将空间设计成室内的风景,家具则若隐若现地隐没于风景之中。每一件独特的家具中都体现着景观学的理念,每一件家具都是一处"风景",每处风景间既能够与彼此相呼应,也能与彼此相分离,它们能脱离环境成为独立的物体。作为风景的一部分的家具——桌子和展示柜——是用一层层的木板制造的,其上罩着一层玻璃。如同一件七巧板拼图玩具,桌子和椅子可以拆开组成不同的造型。

罗斯柴尔德银行总部和家具(ROTHSCHILD BANK HEADQUARTERS AND FURNITURE)

英国,伦敦,1999年

皇家宫殿酒店与赌场(ROYAL PALACE HOTEL AND CASINO)
瑞士,卢加诺,1999年

卢加诺的冰川地貌成就了流畅起伏的风景,而一座城市就建于冰川地貌之上。我们的景观选址范围始于山巅,景色如岩浆流一般从山巅倾泻下来,向着卢加诺湖流去,而酒店的正面则阻拦了一部分倾泻的风景。酒店和赌场设施被容纳并深陷于风景之中,使露天的风景和庭院增添了一束束的光线,营造出了不一样的氛围。酒店被置于已有的建筑之中,与已有的立面相连,而赌场则位于远离酒店的地方,处于山势较低的一端,使建筑的立面保持着古老的形态。

大都市,沙勒罗伊当斯舞蹈艺术中心(METAPOLIS, CHARLEROI DANSES)
比利时,沙勒罗伊,1999年

弗雷德里克·弗拉芒(Frédéric Flamand)为沙勒罗伊当斯舞蹈艺术中心(Charleroi Danses)所设计的舞蹈作品唤醒了这座城市的韵律。景观由多层编织结构构成,其材料多种多样,塑造出了流动的杂糅空间,使之与舞者的姿态相吻合。弗拉芒的舞台舞蹈跟随着并激活了空间的形态变换,将舞者卷入空间的复杂之中,继而利用空间俘获并解放他们。结构本身一直处于不断地变换之中,以一系列的挤压和释放不断地展现着自己。三座跨度10米的透明桥梁形态各异,当舞者在它们之间穿梭的时候,它们可以变换成各种各样的造型。通过道具的过渡和灵活的结构,将舞蹈拓展成了四维的模式。

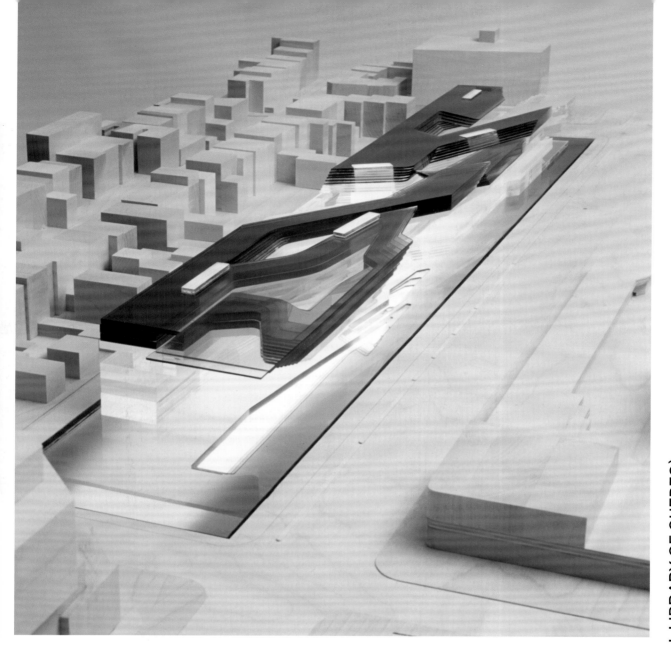

魁北克国家图书馆（NATIONAL LIBRARY OF QUEBEC）

加拿大，魁北克，蒙特利尔，2000年
与帕特里克·舒马赫合作

我们为这座新图书馆设计了各种各样的服务功能，比如珍藏的手稿和24小时的开放区。这项设计为诸如此类的功能区营造了自成体系的空间，并将其在环境上与图书馆的其他部分融为一体。通过区分这些"功能区"，我们颠覆了系统上的差异化，使大型公共建筑不再受制于分区间的不一致。主要的建筑理念是要营造出不断延展的空间，引导着我们进入新的景象之中，然而却遵循了分化的逻辑，将其区分为符合规律的分支空间——就像知识树一般——分支空间如同藤蔓一样侵入巨大的建筑实体。游客们可以沿着藤蔓的分支进入建筑的上部，进而选择进入藏书或是阅览室。主要的藏书馆被塑造成了梯田样式的山谷状空间，书籍被一排排地陈列于藏书馆周边，而阅览区则位于藏书馆的中央。

伯吉塞尔滑雪台 (BERGISEL SKI JUMP)

奥地利,因斯布鲁克,1999—2002年

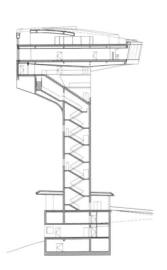

作为奥运会场馆大型整修的一部分,在设计位于提洛尔的伯吉塞尔山上的滑雪台时,我们将高度专业的滑雪设施与公共空间融合在了一起,其周围环境中建有一座咖啡厅以及一座观景台。这些不同的设计被组合为统一的形式,从而将斜坡的流势一直延伸至天际深处。这项设计长约90米,高约50米,它由一座塔和一架桥组合而成。在结构上,它被分成了一座垂直的混凝土高塔以及一座绿色的建筑,其中包括坡道和咖啡厅。两架电梯能够将游客带往咖啡厅,在那里他们可以欣赏阿尔卑斯山的风景,以及其下的运动员如何在因斯布鲁克的天际线之间飞跃。

费诺科学中心（PHÆNO SCIENCE CENTRE）

德国，沃尔夫斯堡，2000—2005年
与克里斯托斯·帕萨斯（CHRISTOS PASSAS）合作

费诺科学中心在德国算是比较异类的建筑，试图展现给游客一定程度的复杂陌生感。这座建筑坐落在一系列由阿尔瓦·阿尔托（Alvar Aalto）、汉斯·夏隆（Hans Scharoun）以及彼得·施韦格（Peter Schweger）所设计的文化建筑周围，它那立于地面之上的庞大体积看起来疏松多孔，这座建筑与大众汽车城相连，与周围大规模的建筑相得益彰。

这栋建筑是按照特殊的体积结构逻辑设计的。楼层并没有一层层地相叠，大厅之上也没有搭建巨大的屋顶。从建筑内探出的圆锥体以及插入建筑内的圆锥体支撑并构造了其上巨大的建筑体，于是，在圆锥体的上方顶立着一个盒子一般的建筑结构。通过圆锥体中的通道，人们可以到达盒状结构的内部，而其他的通道则被用于照明内部空间以及容纳服务设施。圆锥体的构造依据了城市轴的规划原则：一个圆锥体被用作主入口，另一个作为礼堂，另外三个汇聚成了展览空间，位于大厅之下。公共桥梁如虫洞一般在建筑的内部穿梭，它将建筑内外的空间融为相互贯通的整体。

给人耳目一新的感觉，融合为相互贯通的整体——在选材上，我们依旧遵循这两个原则。我们选择了光滑多孔的降噪材料，以方便投射灯光，便于人们观察建筑中的活动。我们在材料之下铺了一层照明设备，折射的灯光便照亮了底面，它将斜切的空间蒙上了一层别样的氛围。光与影形成了可以左右视觉的系统。我们选用了加固的混凝土建材，因为它们易于塑造出流线型一样的自由造型。地面以及屋顶的结构由双向网格板构成，工作区、礼堂、主入口以及行政区均由抗震墙支撑。为了最大限度地重复模板和加固建筑，平行四边形的网格板被设计呈棋盘式分布，于是楼盖之间的混凝土板在相交处构成了锐角，以适用于这座建筑的视轴线。

费诺科学中心

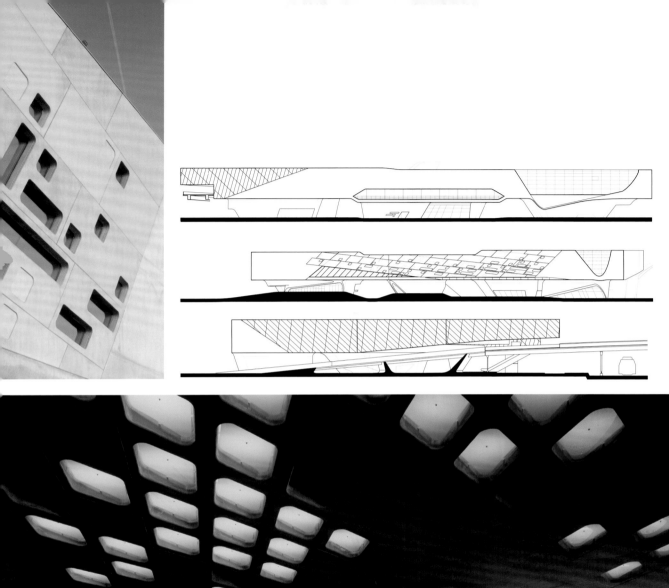

网(MESHWORKS)
意大利,罗马,2000年

在罗马的美第奇别墅(Villa Medici),每两年都会举行一次当代艺术展。我们要与40位建筑家和艺术家一起改造花园的一部分,使其与这座城市融为一体。于是,我们将网状构造投向长达9米的罗马古墙上,从而构造出了典型的文艺复兴风的花园广场,为前来参观的游客带来了全新的体验。

蛇形画廊坐落于伦敦的海德公园中。我们负责为其13周年庆典活动设计一个展棚。该展棚位于画廊外的草地上,三角状的顶棚由传统的可拉伸布料构成。在设计场地上,折叠板向着地面延伸,在不同的衔接点与地面相连,从而制造出起伏的效果,以营造出多样的内部空间。在场地内部,布满了设计独特的桌子,其颜色从白色渐变至黑色,从而提升了空间中的变化感。虽然这座造型起初只打算保持五天的时间,但是在公众的要求下,其寿命被延长至两个月。

蛇形画廊展棚(SERPENTINE GALLERY PAVILION)
英国,伦敦,2000年

在2000年的威尼斯双年展上,英国馆以一系列的现代设计而著称。在这项设计中,我们应用了丝带元素:它们被折叠、缠绕、扎起或者是撕开。在四种造型中,有三种是由法兰绒构成的纽带,它们舒展出不同的轨道;第四种造型依旧突出了运动和轨道,从而让游客们在"思想"的故事里遨游。

英国馆,威尼斯双年展(BRITISH PAVILION, VENICE BIENNALE)
意大利,威尼斯,2000年

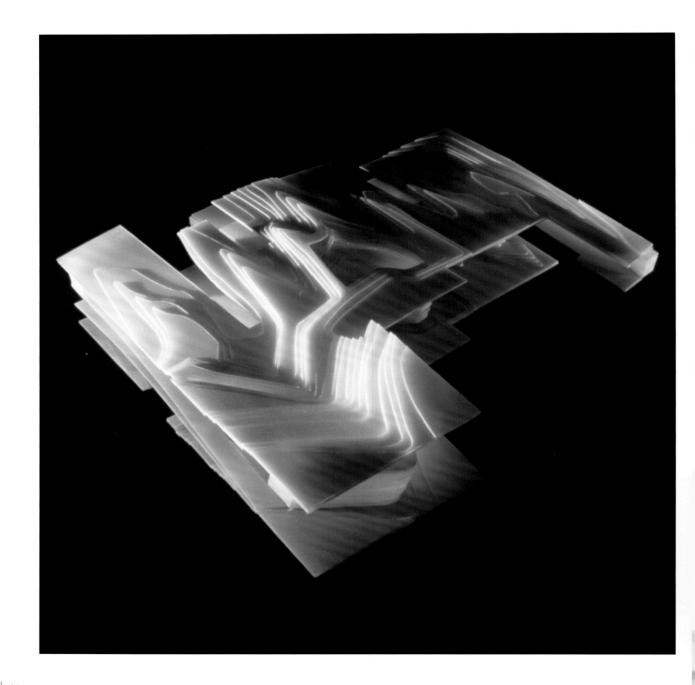

格拉茨美术馆（KUNSTHAUS GRAZ）

奥地利，格拉茨，2000年
与帕特里克·舒马赫合作

在这项设计背后有一个决定性因素，那便是意在让这座博物馆跨越伦德凯大街，穿街而过向着河岸延伸。为了达到较好的采光效果，悬臂结构要足够高。它看起来像是一座高高的天篷，3~6米高，覆盖了极具延展性的空间，使其成为一处宽敞的公共活动场所。天篷的形态遵循蘑菇柱的构造，其顶端部分逐渐变细，由加固混凝土构成。美术馆位于最长的悬臂结构之下，其主垂直结构绕着最大蘑菇形构造旋转一周，围绕出中空的空间。张开的翅状结构在建筑的上方展开，从而在结构的上方营造出巨大的张力，与结构的下部压力达成一种平衡。天篷之内是密闭的空间，这样的设计适应了讲座以及表演的需要，为建筑配备了可选择的隐秘、隔声和避光的效果。

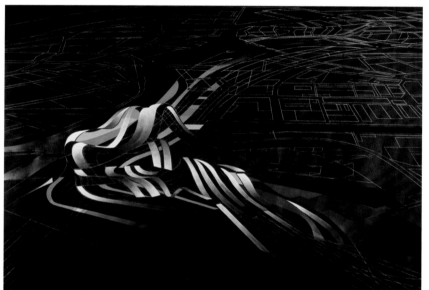

斯特拉斯堡大清真寺（LA GRANDE MOSQUÉE DE STRASBOURG）

法国，斯特拉斯堡，2000年
与帕特里克·舒马赫合作

我们设计这座新清真寺的理念是将其建于一个矩阵之中，矩阵的一边沿祈祷的方向蔓延（也就是朝向麦加方向），另一边则沿着莱茵河的弧度延伸。而在这两条轴线相交的地方，矩阵向两侧展开，滋生出成片的建筑。建筑开始向两边蔓延的相交处，便是这座清真寺所在的位置。在这项设计中，这座建筑的各个功能区被清晰地分开：寺外的空间与街道齐平，而清真寺与后院的位置则高于整个城市的地面。底层较低的世俗区被用作入口，由此可以进入中心院落，也就是冥想区。公共区域的建造方式遵从了传统的伊斯兰建筑风格，其成排的祈祷室全部朝向麦加，水渠则从建筑底层以及院落中穿过，其几何样式以协调的比例达到了某种平衡。该建筑结构中的流线设计隐喻着伊斯兰的书法造型。

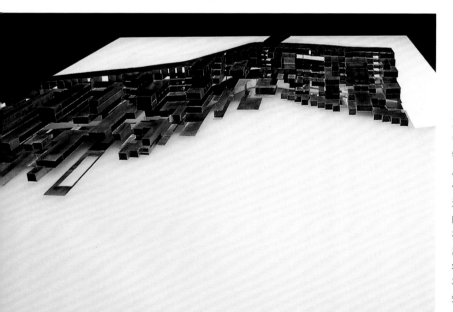

中心JVC酒店（CENTRO JVC HOTEL）

墨西哥，瓜达拉哈拉，2000年

位于瓜达拉哈拉的这家酒店为10座相互交织的建筑提供了背景，其中包括会展中心、博物馆、办公楼、娱乐中心、综合购物大厦和大学校园。我们计划在这里建一座五星级酒店，它位于该区域的北部边缘，俯瞰着一处人工湖。基于酒店房间的构造，酒店的远景以及酒店与其他建筑之间的关系将被利用，以构造出系统的棋盘式布局。作为空间子实体，酒店的房间构造出酒店建筑的整体结构，一层"酒店粒子"状的结构分布于湖泊之上。这座建筑从水平面开始向上延展，一直延展到垂直的建筑立面。该建筑结构从地平面开始向上延伸，构成了当地风景的一部分。

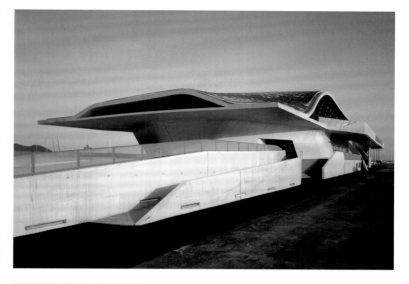

萨勒诺海运码头（SALERNO MARITIME TERMINAL）

意大利，萨勒诺，2000—2016年
与帕特里克·舒马赫合作

这座新的海运码头使城市与海滨之间的关系变得更为亲密。它的形状就像是一只牡蛎，这座建筑坚硬的外壳包裹着其内部柔软的、流动的元素，波状的混凝土屋顶使乘客们免受地中海烈日的曝晒。这座综合建筑由三处相互交织的主元素构成：一处是国际邮轮及游艇码头，另一处是当地及区域邮轮码头，还有一处是行政办公室。作为一个整体，这座建筑将陆地和海洋之间平稳地过渡，它构成了从固态到液态不断变换的地势。码头的内部结构引导着游客们穿过这动态的空间，在空间的四周设有餐馆和候船室等场所。从建筑的窗户和平台上能够看到阿玛尔菲海岸、萨勒诺湾和奇伦托壮观的风景；在夜间，这座建筑则如同灯塔一般在港口处闪耀着光芒——隐喻着此处的诺曼人和撒拉森人的历史。

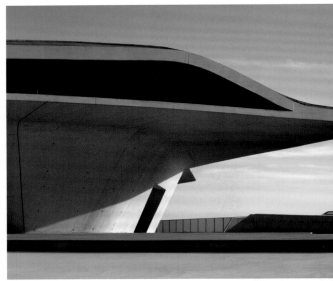

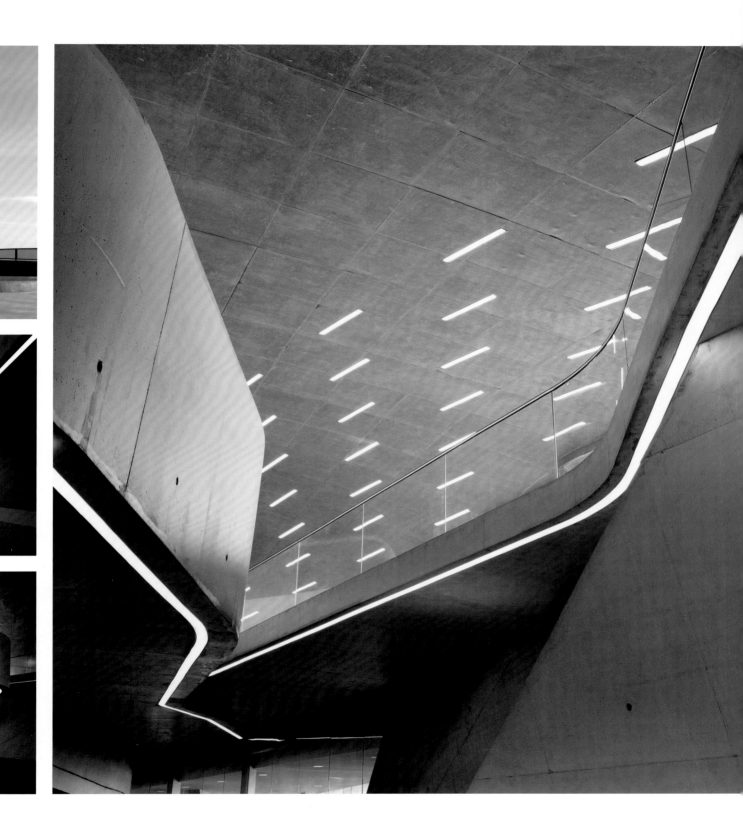

扎哈·哈迪德休息室（ZAHA HADID LOUNGE）

德国，沃尔夫斯堡，2001年

这座沃尔夫斯堡艺术博物馆内有两个多余的区域一直没有被利用，我们想要将空间设计成多功能的休息区。作为费诺科学中心（第116页至第121页）项目的前驱性实验设计，这片休息区中包括会晤室、电影礼堂、等候室、会客室、公共广场、画廊和大厅。颇具活力的流动空间将底层与上层的空间贯通起来，而"休息室"这个词使人直接联想到与外界隔离，且舒适的感觉。安静与热闹，放松与刺激，这一切都在空间中相融合。

阿尔贝蒂娜博物馆扩建（ALBERTINA EXTENSION）

奥地利，维也纳，2001年
与帕特里克·舒马赫合作

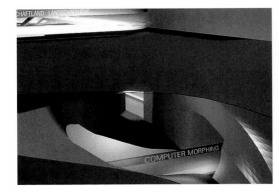

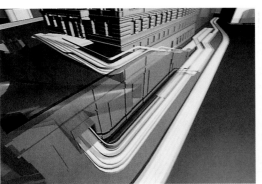

在扩建这座博物馆时，我们用图像体现结构，编织动感的线条，视觉上将周遭与原有的基底相融合。我们选用了两种不同的介质，以使这些结构变得更为形象化。首先，流线型的钢筋以及玻璃构成了线性的网，它被置于延展的路面上，一直摆荡至主入口处。其次，三个对角相连的支撑锥体如同闪电或观光隧道一般，刺穿了堡垒的巨大外壳。阿尔布雷特·丢勒（Albrecht Dürer）的投射几何图形及其在巴洛克建筑中的应用，提供了这个射线状结构的创意。

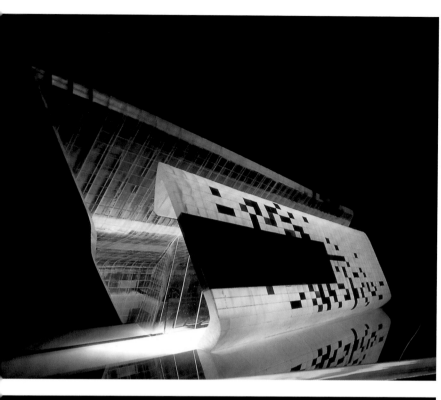

东京古根海姆临时博物馆（TEMPORARY MUSEUM, GUGGENHEIM TOKYO）
日本，东京，2001年

台场岛是一处进行文化实验的完美地点，动态的城市空间分布于综合用地之上。经过长达10年的政府干预，临时的古根海姆博物馆如今已经成了推动文化发展的催化剂。为了符合该建筑架构的临时性，我们使用了重量较轻的材料。两面像纸一样薄的折叠板相搭在一起，覆盖着宽敞的空间。它们不仅构造出标志性的造型，还塑造了可以被持续重新定义的空间。虽然空间概念极其简单，但是却把折叠板的大小和抽象的程度以及动态的造型组合在一起，并且被赋予了蛇皮一样的外表，营造出令人振奋的空间效果。

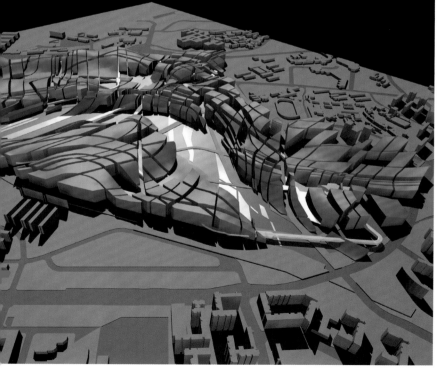

一北总体规划（ONE-NORTH MASTERPLAN）
新加坡，2002年
与帕特里克·舒马赫合作

新加坡总体规划为我们提供了一个机会，让我们首次在整个城市地区探索人工风景构造。这一大胆的举动营造出了惊人的效果：天然的天际线、极具辨识力的城市全景、高密度的广场和街巷，这一切都营造出了很强的场地感。沙丘一般的巨构形式赋予了空间以连贯性，屋顶表面也置入了构造之中，更是提升了连贯的效果，如今这样的造型在现代都市中已是罕见。与此同时，诸多建筑体——高的、矮的、宽的、窄的——在组合力的作用下融合在一起。街道以及路线被塑造成起伏的流线型，从而赋予形状以强烈的一致性，并在一致中掺杂了无限的多样性。由此，建筑又具备了"浑然天成"的优势，而非不切实际的几何图形。

宝马工厂中央大楼（BMW PLANT CENTRAL BUILDING）

德国，莱比锡城，2001—2005年
与帕特里克·舒马赫合作

作为整座宝马工厂综合建筑的活动中枢，所有的活动都在这座大楼中聚集又分散。这座大楼需要容纳生产线、工人和来参观的游客，他们穿梭在整个空间之中。在动态的空间系统中，这座建筑的核心地位是显而易见的，它占据了工厂的整片北部区域，并成为位于汇合点的中心建筑。这一核心区域——或者也可以称为"商业中心地"——为员工和行政服务部门提供了交流的场所。

从结构上来讲，这座建筑的基本构造是一处剪刀状造型，将地面与一楼协调地连接起来。两排梯形板被搭成了巨大的阶梯造型，一层一层地由北向南排列，另一排则反向由南向北排列，于是两排梯形板之间形成了中空的空间。空间的底部是审核区，这里吸引了每个人的注意力。在中空的空间之上，展露在我们面前的是半成品汽车，它们在工作间之间的轨道上移动着。宽阔的梯形层板足以提供灵活的模式，因而较之于单层平板，能为我们提供更多的视觉体验。而内部结构的透明性更是增添了视觉体验。混合的功能区替代了传统的各类群体的隔离，比如，工程办公室与管理办公室就置于工人劳动的空间之中，日复一日近距离指导工人。

我们将大型停车场设置于建筑右侧，这样就避免了将过大的停车场置于正前方。交通工具在建筑内外穿梭的动态融入停车空间的构造之中。由此，整个场地便都处于动态之中，色泽和光晕在场地的轮廓之上变换，并在建筑的内部达到最佳效果。

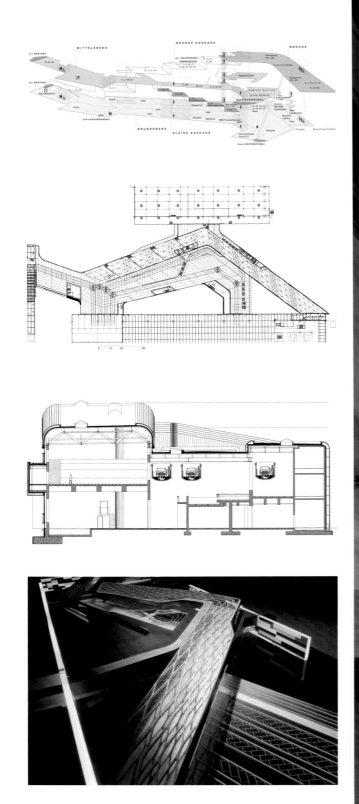

　　奥德罗普格博物馆的扩建恰巧给我们提供了一个机会，让我们能够把不同的建筑和花园组合成一个整体。扩建部分不仅为建筑场地的景色增添了离散性的特点，而且还为花园的地势增添了几分色彩。将位于建筑前的平坦地带与后部的斜坡花园一分为二，建筑的轮廓被精心地抽象化，然后被提升并曲折成了贝壳状，从而构造出这座博物馆的轮廓。游客们在花园中与艺术不期而遇，他们向着扩建部分走来，站在地面不同的制高点上可以看见不同规模的作品。博物馆内部的风景被构造成了一系列流动的串联空间，从而与外部空间形成极具张力的对比。

　　原有的法国画廊将这座新的建筑物与一座院落分开。画廊位于前厅和购物区之后，以从南向北的方向伸展开来。而电梯、楼梯以及存储区的后墙则将大厅与画廊区域一分为二。一道斜坡顺势将临时画廊空间与永久画廊空间一分为二，将游客们带往面向花园的多功能厅和咖啡厅。在博物馆综合设施中心的丹麦和法国藏品展览长廊，高更（Gauguin）画廊位于花园的尽头，一条新建的画廊与其相连，将一座窄小的后院与其隔开。在所设计的建筑物内，这一新的空间是面积最大的一处独立区域，从而留足了余地与其他画廊在视觉上相衔接，并且与室外花园中回廊一般的空间连为一体。

奥德罗普格博物馆扩建（ORDRUPGAARD MUSEUM EXTENSION）
丹麦，哥本哈根，2001—2005年

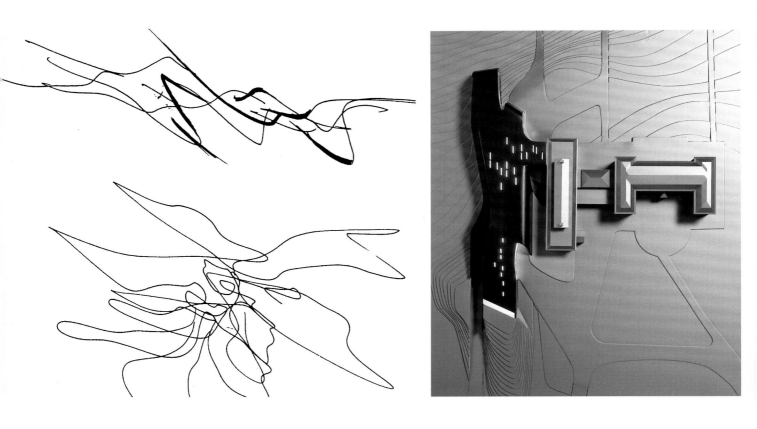
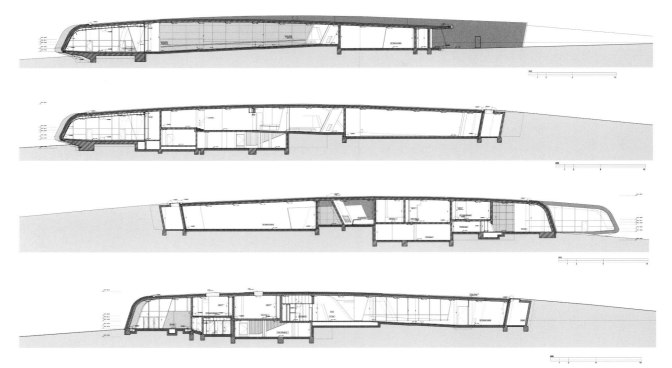

奥德罗普格博物馆扩建

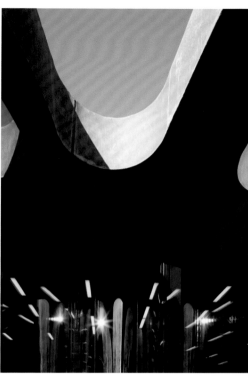

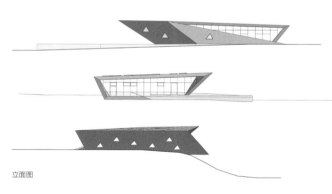

立面图

　　这项设计的意图是将丰富而茂盛的大自然环境过渡至人类建造的停车场与医院。我们以同一种材料——特种高强度钢覆盖了可视的屋顶以及两面相对的墙。这种建筑造型由折叠的表面和相接的地基板构成。屋顶巨大的垂悬结构将整座建筑物与地面相接，混凝土基座则将建筑的中心部位与周围的景区相连。建筑内部的房间围绕着开敞式的厨房布局，办公室则坐落于北部地势偏高的位置，这里便于俯瞰停车场和入口。朝东的房间更为隐秘，从外部看，它们构成了一面半透明的立面。站在东北部的入口处，视线可以直接穿过中间的厨房空间，望见朝向南边的玻璃立面造型。

美琪第五癌症中心（MAGGIE'S CENTRE FIFE）
英国，柯科迪，2001—2006年

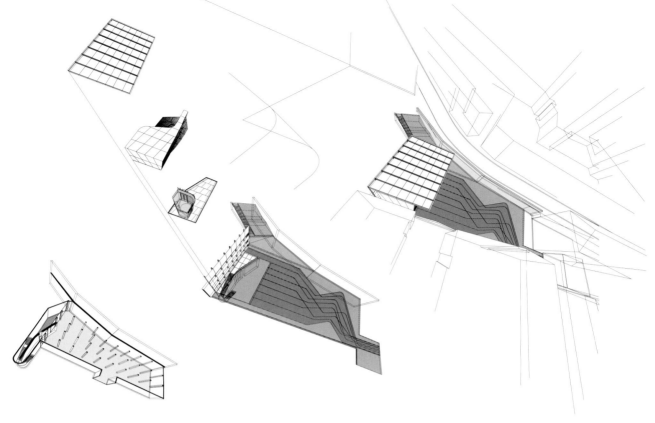

洛佩兹·雷迪亚酒庄（LÓPEZ DE HEREDIA PAVILION）
西班牙，哈罗拉里奥哈，2001—2006年

这位客户找到并要求我们设计一座"世纪之末（Fin de siècle style）"风格的酒庄。新酒庄的内部将保留着另一座老酒庄，内部的这座老酒庄最初是由客户的祖父于1910年建造的。置于新的建筑之中，这座老酒庄宛若一颗珠宝。如同俄罗斯套娃一般，它最终将被嵌套于新酒庄及其延展部分的内部。这样，新酒庄外部就又覆盖了一层延展构造。经过多次研究之后，我们设计了一座横截面般的新酒庄。老酒庄外部嵌套的建筑造型原本为矩形结构，截面却将其扭曲为变形的形状，使它看起来很像一个玻璃酒瓶。我们并非有意制造出这样的效果，但它确实让人过目难忘。我们将老酒装进了新瓶子里。

专题展览 (MONOGRAPHIC EXHIBITION)

意大利，罗马，2002年
与帕特里克·舒马赫合作

为了庆祝国立二十一世纪艺术博物馆（第106页至第109页）的建设，国家当代艺术中心举办了一届展览。这届专题展览构建于原有的维亚·圭多·雷尼（Via Guido Reni）建筑之内。展览的内容以国立二十一世纪艺术博物馆的流线型造型为出发点，所展出的设计作品覆盖了各个方面，它们彼此呼应，相互交织，以流动的形式交换着理念与设计灵感。一面蜿蜒的墙俨然而立，与当代艺术中心展览的那面墙几乎如出一辙，将展品贯穿并连贯起来。

普莱斯塔楼艺术中心 (PRICE TOWER ARTS CENTER)

美国，俄克拉荷马州，巴特尔斯维尔，2002年
与帕特里克·舒马赫合作

在弗兰克·劳埃德·赖特所设计的普莱斯塔楼附近，我们将设计一座新的艺术场所。动态的造型有的被高高悬置，有的棱角分明，有的则被建得很低，从而使整座建筑进入了大胆的对话之中。通过仔细地研究城市以及自然地貌，城市网格布局以及城市的整体流势，我们将这项设计叠加于塔楼所定位的倾斜轴上，并使其层叠于赖特那座"织物纹样砌块风格"（textile-block）建筑的倾斜轴上。最终，我们营造出了一种效果，从场地的鲜活图案里，这座建筑脱颖而出，这也恰恰契合了赖特本身的建筑理念。

万塔之城，威尼斯双年展 (CITY OF TOWERS, VENICE BIENNALE)

意大利，威尼斯，2002年
与帕特里克·舒马赫合作

纽约的世贸中心被摧毁了，这是一个悲剧。这向我们提出了一个问题，我们该建造一座怎样的建筑来取代它呢？与其建造一座寓意深刻的建筑以回应此悲剧事件，我们不如选择扪心自问：什么样的组织结构能够满足当代商务生活的需求，什么样的语言能表达这种生活？现代都市的功能和美学是什么？如果可能的话，如何将其本质展现于曼哈顿的天际线之间？

我们所设计的这座政府综合大楼包含一间档案馆、一间图书馆和一套运动设施。虽然这座建筑严格地沿着设计线延展，但其形状却类似于一根巨大的树干，它被水平地放置于地面上。档案馆坐落于树干最稳固的底部，树干中部为更为通透的图书馆，而运动设施以及灯火通明的办公室则位于树干的顶部，树干在顶部分叉为更为轻巧的结构。整根树干伸出了多个入口，使人能够进入不同的机构之中。在底层，不同的机构共享着公共的空间。而在上层之中，每一个机构都是独立的，其内部设有贯通点，以用于内部纵向的来往。

刚一抵达这座建筑，游客们便可以穿过大厅抵达底层档案馆的教育空间。或者也可以乘电梯或扶梯抵达一层的主图书馆艺术区。建筑的一侧建有凹陷的玻璃走廊，这里被用作艺术区，从此处既可以直接抵达图书馆的阅览室，也可以前往档案馆的阅览室。艺术区位于整座建筑的核心部位，三座主要的公共设施位于艺术区的中心，三个机构共享着这三座公共设施：一间礼堂和两间会议室。这些共享公共空间同样也构成了建筑的中心结构，它们从树干上凸出来，为路人们形成了悬置的宏大天篷。

皮埃尔·韦弗斯政府大楼（PIERRES VIVES）
法国，蒙彼利埃，2002—2012年

皮埃尔·韦弗斯政府大楼

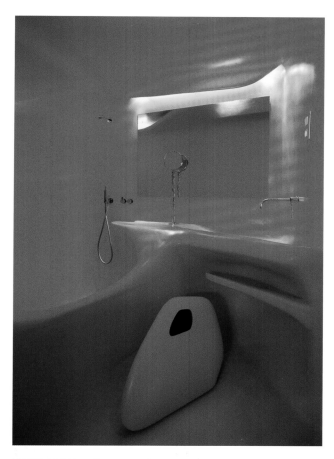

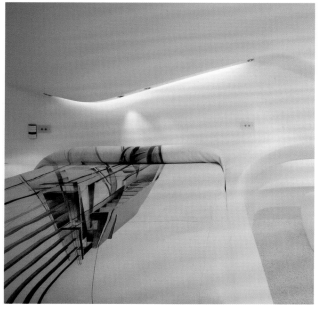

布兰德霍斯特博物馆 (MUSEUM BRANDHORST)

德国,慕尼黑,2002年
与帕特里克·舒马赫合作

这座建筑位于慕尼黑博物馆林立的地区。该设计注重将建筑整体与城市环境天衣无缝地融为一体。一道斜切口如同峡谷一般打开了这个"艺术盒子",由此在建筑内部营造出连贯的垂直循环空间,从而构造出一道斜角走廊,打开了从特蕾莎大街与土耳其大街拐角处和现代美术馆之间的斜视角。在建筑内部,建有公共大厅、接待室和博物馆商店,给人以别具一格的空间体验。建筑上下均有内部通道,它们沿着斜切口的两面墙伸入到展览馆内部,使高度各异的展馆空间贯通为一体。在建筑的中心区域,展品那错综复杂的美感尽情地展现在人们眼前,赋予了展馆空间以斑斓的多样性。

佐罗扎伊半岛总体规划（ZORROZAURRE MASTERPLAN）

西班牙，毕尔巴鄂，2003年至今
与帕特里克·舒马赫合作

纳文河气势磅礴地流过这片区域，在拐角处形成大面积的河滩，从而巧妙地影响了毕尔巴鄂精制的城市棋盘布局。我们这项规划所针对的对象为前港口和工业区，在设计时充分考虑了纳文河的流势。沿着狭长的场地，我们所设计的线性建筑群——铺开，营造出构造精致的建筑区域。这些建筑顺应了原有建筑布局较小的格局，并向外扩展与更广阔的空间相贯通。以这样一种方式，这项设计既符合了旧建筑的格局，又融入了新建筑的元素，从而使二者均面向广阔的公共滨水空间。

为数众多的桥梁以及延展的有轨电车系统将城市连为一体，从而使下游的社区与市中心相连。一系列面积超过1000平方米的建筑街区构成了优雅的建筑群，从而赋予了设计规划以整体的统一性。街区的地面高度恰好可以防洪，并为其下的地下停车场提供了空间。地面随着拓展的建筑群而不断延展，缩短了河滨步行道与河岸之间的距离，让当地居民与河流进行更为亲密的接触。

地面高度的微妙变化赋予了公共和私人空间以丰富的纹理，从而在私人与公共生活之间达成了微妙的平衡。在总体结构中，质地紧密的环境轻易地打造出来，以配合布局上强烈的孔隙感。由此，未来的居民和工人们便能享受丰富多样的户外空间。

英国广播公司音乐中心(BBC MUSIC CENTRE)

英国,伦敦,2003年
与帕特里克·舒马赫合作

我们设计了一座新的英国广播公司音乐中心,它构成了一座规模更大的综合设施的公共中心建筑,路人们都被吸引至这座建筑那高高的遮蔽之下。配套设施包围了工作室,呈"带状"分布的房间分割了表演空间,从而构成了不同音响效果的子循环系统。按环状布局的阳台全部分配给了交响乐表演人员、观众和技术人员,激活了排练空间和工作室录音的空间。

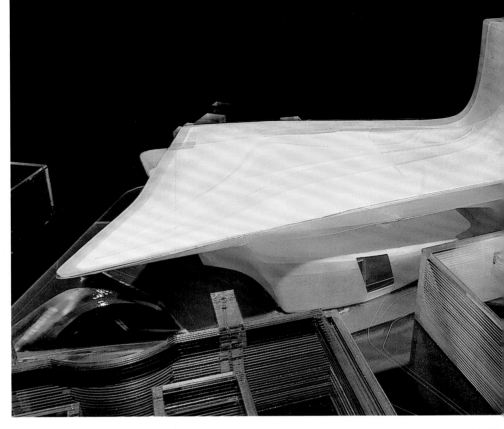

欲望(DESIRE)

奥地利,格拉茨,2003年
与帕特里克·舒马赫合作

我们为格拉茨的施泰尔秋季艺术节设计了这项作品。"欲望"是贝艾特·福瑞(Beat Furrer)所创造的一部当代歌剧,通过讲述俄尔普斯(Orpheus)与欧律狄刻(Eurydice)之间的故事,突出了心理与情感层面的变质。我们将舞台设计为可感知的风景,按照棋盘式布局将舞台扭曲为一系列动态的部分,使它们随着叙事的进程和音乐的节奏而不断变换。指向性很强的线段在布景中曲折穿过,风景则被劈开,一分为二地呈现在观众眼前。内部与外部、上部与下部、配乐与舞台之间的界限被模糊了。

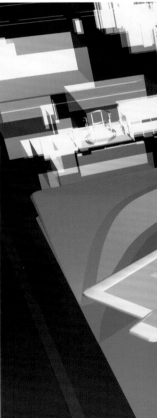

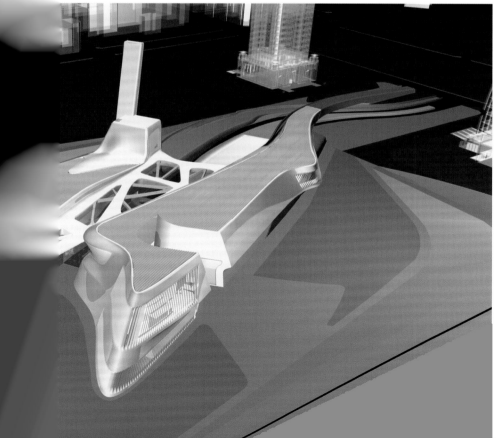

台中古根海姆博物馆（GUGGENHEIM MUSEUM TAICHUNG）

中国，台中，2003年
与帕特里克·舒马赫合作

　　我们在设计这座博物馆时，将其看作是不断变换的活动空间。大规模的动力要素构成了"舞台的机械构造"，从而使得展览空间能够完成大的转换。这些转换本身即为城市风景中的公共奇观，从建筑外部便能看见这样的壮景。各种各样的斜面贯穿于建筑之中，形成了循环的系统，从而使大众可以自由地穿梭。可移动楼层、墙壁和平台则塑造了诸多内部空间造型。

雪之秀 (THE SNOW SHOW)
芬兰,拉普兰,2003—2004年
与帕特里克·舒马赫合作

冰与雪是让人赞叹的自然元素,它们那流动的造型以及连贯的构造总是让人吃惊。我们与艺术家蔡国强一起创作了这一处风景造型,它不仅深化了冰雪元素所塑造的氛围,而且还表达了元素从一种状态至另一种状态的转化。两处彼此呼应的风景映入了眼帘:一处由雪元素构成,另一处由冰元素构成。流动的空间和天篷将游客们容纳至流光溢彩的冰川之间,墙壁向上卷曲成为天花板,进而延展为反重力的结构。灯光投落至空间之中,宛若光线的"脉络"交织于结构之中。在风景之中,蔡国强设计了虚拟的火焰场景,火焰如瀑布一般形成了蓝色的溪流、池塘与阶梯,从而重新塑造了结构的形状,并将这一区域转换为人造的熔融空间。剩余的冰则被置于自然之中。

2012年纽约奥运村 (NYC 2012 OLYMPIC VILLAGE)
美国,纽约,2004年
与帕特里克·舒马赫合作

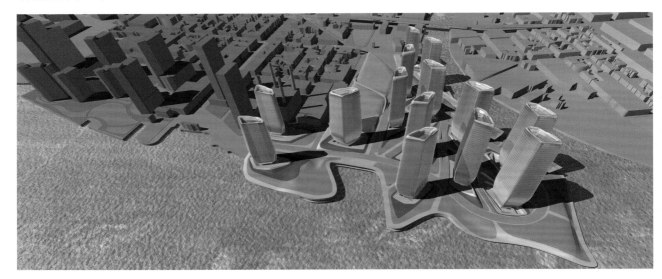

我们的这项总体规划设计打开了亨特角(Hunters Point)的未来,使其在纽约东河上迅速地发展。我们为纽约2012年奥运会申办活动设计了这一座奥运村,它由一系列的塔楼构成,其占地面积不大,却通往宽敞的独特景观。在此举目远望,游客们能够看到市中心和下曼哈顿区的壮景。起伏的地势抬高了塔楼垂直面以及水平面,赋予了设计进程以最大的灵活性。底层的层叠、分割、照明策略与艺术品的融合进一步界定了街道和内部轴线。

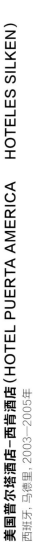

美国普尔塔酒店-西肯酒店（HOTEL PUERTA AMERICA　HOTELES SILKEN）

西班牙，马德里，2003—2005年

　　我们被告之为酒店设计一层楼，包括30间客房和所有的公共空间，并且我们拥有所有的决定权。我们的设计目标是要构建出动态的自定义空间，从而最大限度地提供流动自如的空间以及浑然一体的体验。我们利用这个机会创造新的建筑内部语言，而数码设计以及设计能力的提高为我们提供了便利。这种新的讨论突出了设计本身复杂的动态性质，并将分离的形状和纹理融为一体。

　　不断延展的一体式表层覆盖了房间、地板、墙壁和家具，使它们成为艺术品。每一处独立的元素——墙壁、卧室门以及门上的液晶标志、通往洗手间的滑动门、浴缸和组合式梳妆盥洗盆、床、衣架、椅子和可折叠为桌子的悬臂式长椅——被环绕为单一的弧形布局。卧室套间中更为独特的元素是颜色的使用，顾客们可以选择一间雪山白色或者纯黑色的卧室，他们甚至可以选择一间配有黑色洗手间的白色卧室、一间配有白色洗手间的黑色卧室或者是一间配有橘色洗手间的黑色卧室。

155

广州大剧院的造型如同溪流中的一颗石子，在侵蚀作用下被磨成了圆润的形状。这座建筑坐落于河岸上，处于这座城市文化发展的核心：它那独特的双卵石状设计造型面向珠江，为城市风光增添了一抹靓丽的色彩；在它的作用下，相邻的文化建筑与珠江新城的国际金融塔楼融为连贯的整体。歌剧院观众席能够容纳1800人，剧院中应用了最先进的音响效果，较小的多功能厅（能够容纳400人）被用于观看表演艺术、戏剧和音乐会，观众可以围绕舞台而坐，观看剧场中央的表演。

这项设计的灵感来源于自然风景的概念，其中包含建筑与自然之间妙趣横生的互动，突出了侵蚀、地质和地势的原则。这项设计主要受到了河谷造型的影响，按照侵蚀的效果定义了建筑的造型。当地风景中的折线定义了歌剧院内部的区域和分区，切割出向内和向外的峡谷以塑造环路、大厅和咖啡馆，并使自然光线照射到建筑物深处。互异的元素以及不同的层次之间平稳地过渡，它们继续延续着风景的走势。专门定制的装饰物由玻璃纤维加固的石膏制成，它们被用于歌剧院内部，从而延续了建筑的流线型与流畅性。

广州大剧院（GUANGZHOU OPERA HOUSE）
中国，广州，2003—2010年

广州大剧院

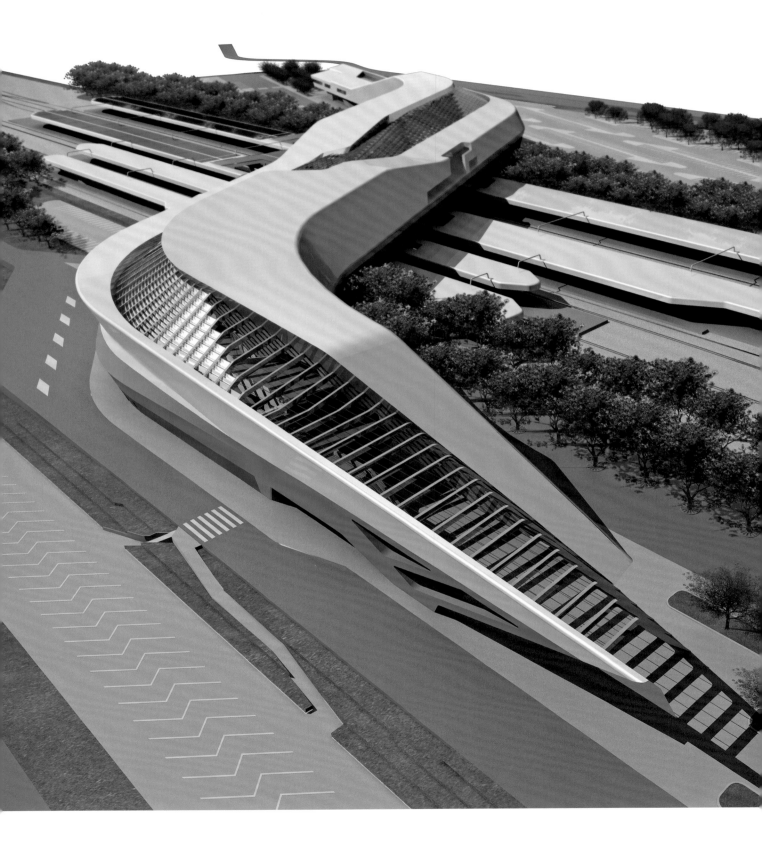

新的火车站在铁轨上方建造了一座桥，这项建筑设计的关键处在于建造一处构造精良的交通转换站，并将其作为通往那不勒斯的门户。桥梁的设计造型意在扩大上部大厅的空间。它将连接各式各样的站台，因而其本身便是一个主候车厅。

作为都市化公共链接，这座大厅桥梁不仅将铁轨轨道贯连为一体，还表达了新式中转站作为即将建成的商业中心是如何将周围的大小城镇相互连接的。在设计理念方面，这座桥梁进一步容纳了两条公园带，它们沿着铁轨运行的方向在建筑场地上延伸开来，从而赋予建筑场地以无限的空间，并将其与周围的风景和商业园连为一体。随着建筑外形的延展，这座建筑物的内部也在诉说着自身的语言，内部的客流轨迹确定了空间的几何学的应用。

阿夫拉戈拉火车站（HIGH-SPEED TRAIN STATION NAPOLI-AFRAGOLA）
意大利，那不勒斯，2003—2017年
与帕特里克·舒马赫合作

加州住宅（CALIFORNIA RESIDENCE）

美国，加利福尼亚，圣地亚哥，2003年至今
与帕特里克·舒马赫合作

这座家庭住宅俯瞰着太平洋，建筑场地上新的综合地貌遮蔽了这座新的建筑结构，一座包含动态顶棚的围篱跨越了私人以及家庭聚集的空间，以突出海上的风景。通过平面图和剖面图上的极线变形，建筑的外形回应着其内部的动态效果，从而将所有的建筑元素融合为天衣无缝的一体，并且使游客能够享受一览无余的风景，使他们能够畅通无阻地进入前厅和内部景致之中。这一单幢建筑与周围的结构相互契合，对比着美国郊野司空见惯的地貌。

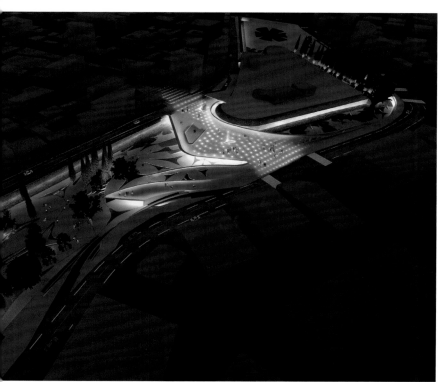

为尼科西亚这座城市设计城市广场时，我们遇到了根深蒂固的历史问题和其他矛盾：虽然威尼斯城墙宏伟的要塞意在保护这座城市免受侵略，但它也将旧城与新城隔开。

我们采用了建筑性干预的设计理念，将其归为规模更大的城市规划的一部分。这项建筑规划意在组织并整合全部城墙，将护城河与现代城市的边缘区融合为统一的整体。在这一宏大的预想中，护城河俨然一条"绿色的带子"——一条环绕着城墙的项链——它不仅成为了放松和娱乐的核心场所，而且还是艺术展览与雕塑作品放置的地方。

护城河被填高并加宽，通过贯连的街道可以抵达护城河，多亏了地下停车场的普及，这些通往内城的新门户才能免受交通拥堵的困扰。城墙恢复了以往的气派，其两旁的人行道外侧种满了棕榈树，灯光也彻夜闪耀着直到天明。这一切都宣布护城河的意义，它隶属于这座浑然一体的新尼科西亚城。

艾弗蕾希娅广场新设计（ELEFTHERIA SQUARE REDESIGN）
塞浦路斯，尼科西亚，2005—2017年
与帕特里克·舒马赫合作

罗德帕克缆车道 (NORDPARK CABLE RAILWAY)

奥地利，因斯布鲁克，2004—2007年

与帕特里克·舒马赫合作

这条缆车道位于蒂罗尔州的阿尔卑斯山脉之中。为了拿到这项设计，我们打败了斯特拉巴格公司。这项设计包含四座新的火车站和一座位于因河之上的斜拉式吊桥，这是我们在因斯布鲁克城的第二项设计（伯吉塞尔滑雪台，第114页至第115页）。

每一座火车站都有其独特的环境、地貌、样式和循环。两种截然相反的元素——"壳"与"影"营造出火车站的空间特性。轻质有机屋顶结构是由双曲玻璃构成的，它"悬浮"于混凝土基座之上，制造出人造风景以表现其内部的动态以及循环。因为我们想以每一座火车站来表达冰川形成的动态语言，诸如溪流是如何冰封于山腰之上的，所以在设计过程中，我们耗费了大量的时间研究自然现象，诸如冰河堆石以及冰川运动。这种语言具备很强的灵活性，所以我们能够按照不同的参数调整壳状的结构，与此同时保持连贯的结构逻辑性。

汽车工业中新的生产方法，包括数控加工、加热成形以及制造技术，使我们能够精确而自动地将电脑设计转换为流线型的建筑结构。

总平面图

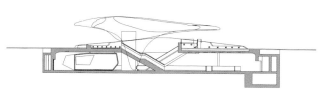

国会站

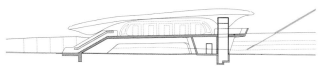

洛温豪斯站

阿尔卑斯山动物园站

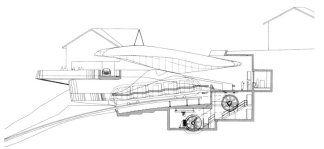

亨格堡站

罗德帕克缆车道

河畔博物馆(RIVERSIDE MUSEUM)

英国,格拉斯哥,2004—2011年

　　格拉斯哥的环境激发我们设计了这座新博物馆,克莱德河为我们提供了历史素材,而这座城市自身则提供了独一无二的特色。这项设计从城区一直延伸至河边,隐喻着博物馆的展区与更广阔环境之间的动态关系,以及二者之间的相互转换,它积极地促成着二者之间的交融。

　　这座建筑内隧道一般的空间,在博物馆的内部形成了一条通道。它构成了城市与河流之间的协调者。这条通道可以与世隔绝,也可以透气,这完全取决于展览的布局。因此,这座博物馆在功能层面以及隐喻层面皆是开放且流动的。博物馆的外部环境和内部设计融为一体,从而将格拉斯哥的历史与未来相贯通。当游客们从一个展区参观至另一个展区的时候,他们会逐渐地深化外在环境带给他们带来的体悟。

　　在这项设计中,我们采用了组合式挤压成型的方式。相对的开口形成于曲折的线段的转折处。横断面的轮廓模仿了压缩的波浪或者"褶皱"。外层"褶皱"围绕出空间,以容纳辅助服务设施和"黑盒子"展览,从而使主要的中心空间内空旷而无柱,以最高水准的灵活性展览这座博物馆中的世界级展品。

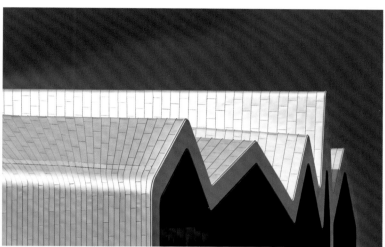

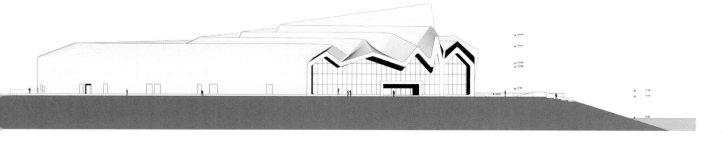

米兰城市生活总体规划（CITY LIFE MILANO）

意大利，米兰，2004—2017年
与帕特里克·舒马赫合作

这项总体规划设计的场地中已有一座大型公园、三幢办公塔楼、零售业建筑、教育与社会设施、一座博物馆以及两座综合住宅建筑，它们都是由不同的建筑师设计的。在这片建筑场地上，以前是一座米兰国际展览中心。作为这座城市的历史性露天广场，这座展览中心位于几条重要城市轴线的交汇处。

现存城市结构的走势激发了我们的灵感，使我们设计出位于核心部位的建筑结构：一座高达170米的办公室塔楼（在建设中），其与两层的零售空间相连，还有一座由七座大楼构成的综合住宅建筑（完工于2014年）。住宅建筑被塑造成了弯曲的环状，一条人行道将其一分为二，使其在视觉和形态上与公园以及忠利塔连为一体。说到建筑的轮廓，这些建筑的屋顶——它们从5层的高度逐渐变换为13层的高度——连成了一条弯曲的线，它呼应着整体综合建筑的流线性造型。建筑围护结构顺应了阳台以及平台的曲线走势，从而塑造了各种各样的私人空间。

若干道路在公园中交汇，形成了一个涡旋状造型，而在造型的中心为一座塔楼。人行道与门店的交汇处由内而外散发着张力，激活了裙房所构成的曲线造型。一股扭力从塔楼向外传送出去——涡旋结构的真正中心——将水平的力量转换为垂直的推力。

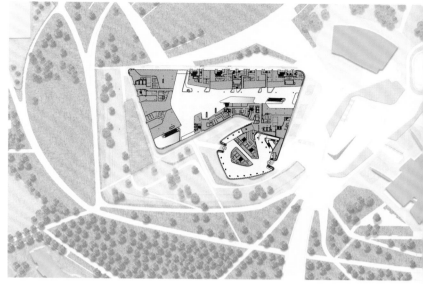

萨拉戈萨桥（ZARAGOZA BRIDGE PAVILION）

西班牙，萨拉戈萨，2005—2008年
与帕特里克·舒马赫合作

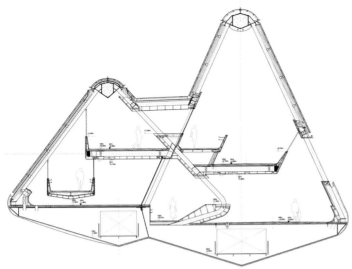

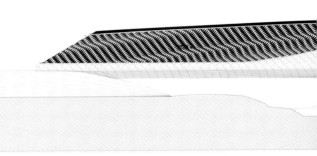

在一条弯曲的道路旁，伫立着一座钻石状的建筑，在研究过该结构之后，我们设计了一座桥亭。它由四个主要的桁架单元构成，也可称其为"纵槽"。它们不仅支撑着结构而且围合了空间。这些纵槽相互堆叠交叉，与彼此相互交织，从而将负荷力分担于四根桁架之间，避免了由单独一根桁架承受所有的负荷。我们不曾预见桁架交织的效果，但它确实适用于我们的设计。游客从一个纵槽游览至另一个纵槽，由此穿过狭小的间隙空间。它们如同过滤层或缓冲带，将声音和视觉体验从一个展览空间传播至另一个展览空间。

在设计这座建筑的外表层时，鲨鱼鳞片为我们提供了迷人的示范，我们将其应用于外表和其性能的设计之中。以简单脊的直线系统为依托，此类样式可以轻易地包裹复杂的弧度，使之在具备表现力和视觉吸引力的同时，达到节约材料的效果。一些简单的面板相互重叠，构成了复杂的样式，由此形成了建筑的外表层。有些面板可绕着中心点合合，形成了建筑正面的入口。光线的质感灵活多变，可以是穿过细小断续空洞的光束，也可以是开阔宏大的光幕。

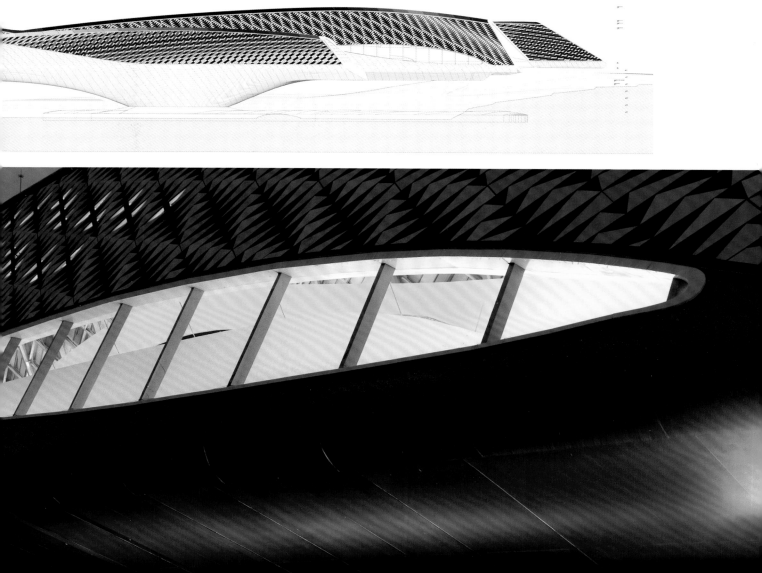

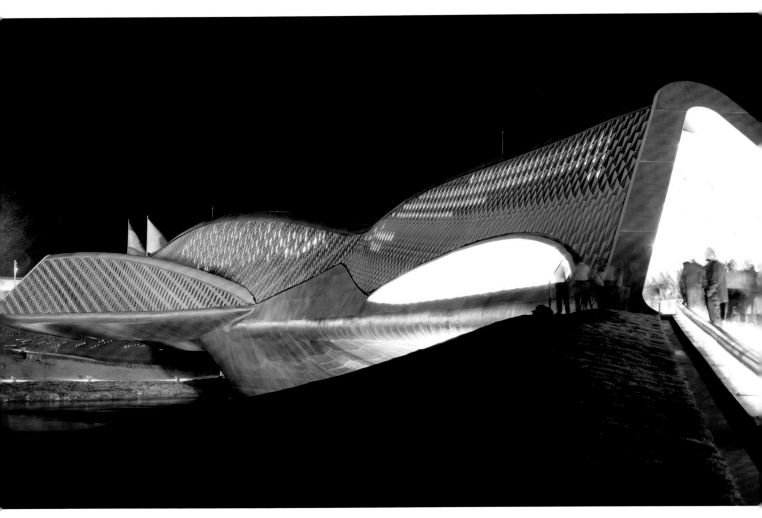

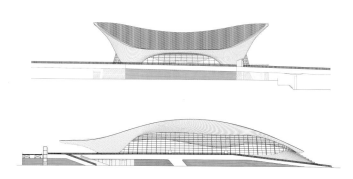

我们为2012年伦敦奥运会设计了这座重要的会场，我们的灵感来源于流动的水流造型。这项设计着重于塑造空间及其周围的环境，使其与奥林匹克园中的河流风景相协调一致。这座建筑位于正交轴上，与斯特拉特福城市桥梁相垂直，其周围分布着三座游泳池：位于大桥之下的一座训练游泳池，位于宽敞的游泳馆内的比赛游泳池和潜水池。游泳馆环绕着桥梁而建，与桥梁连为一体，游泳馆的底部被设计为蹲座墙造型。蹲座墙样式能够将诸多设备容纳于单一的建筑体内，从而使建筑完全与桥梁以及风景融为一体。蹲座墙起于桥梁处，呈瀑布状环绕过游泳馆，止于水道的较低洼处。我们用双曲几何造型装饰抛物线造型的拱形结构，塑造出独具特色的巨大屋顶，它变换起伏着营造出内在的视觉区分效果，以区分竞赛池建筑体和浅水池建筑体。屋顶向外伸展，覆盖过游泳馆基础设施的外层，覆盖着其外的瀑布状造型以及桥梁的入口。屋顶的延伸部分暗示着人们，无论是从瀑布状造型处，还是从桥梁入口处，都可以进入这座水上运动中心。

结构上，屋顶由三处主要的支撑点支撑，屋顶与蹲座墙之间的空隙由玻璃幕墙填满。我们之所以这样设计是为了在比赛期间，能够容纳15 000名观众在游泳池边上，而不必担心结构上出现任何形式的阻碍。比赛之后，我们可以用玻璃面板替换观众席，将建筑内的观众席缩减为2500名，从而将基础场地还原为交流以及精准训练的场所。这座水上运动中心于2014年开始对外开放。

伦敦水上运动中心（LONDON AQUATICS CENTRE）

英国，伦敦，2005—2011年

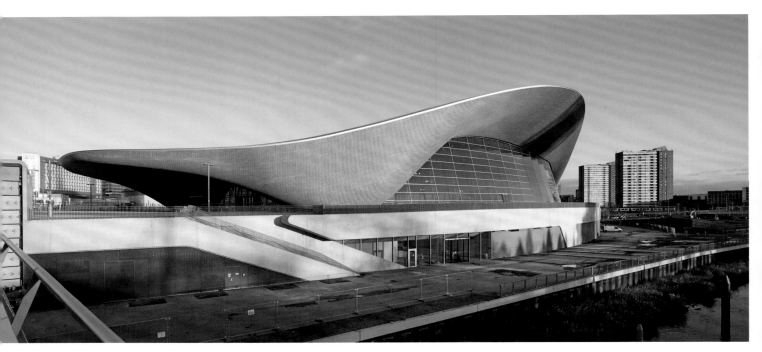

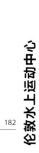

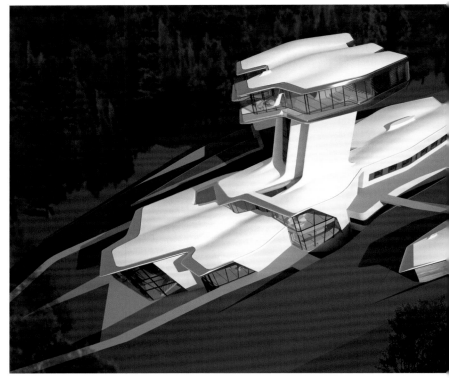

首都山别墅（CAPITAL HILL RESIDENCE）

俄罗斯，莫斯科，2006年至今

与帕特里克·舒马赫合作

这座建筑主要被分成了两部分：在我们的精心设计下，一部分建筑体与斜坡风景融为一体，另一部分建筑体被置于高于地面22米的地方。这座建筑的造型具备动态性，它从风景之中拔地而起，但某些部分仍掩映于山体之间。水平的造型与地形融为一体，人造的阳台占据了原有的地形。外部构造延伸到建筑内部，随后又在周围的环境之中表达并释放着自身。这种相互作用的设计进程消解了内部与外部之间的界限，营造出浑然天成的流动效果，这种效果向着建筑体的高处蔓延，影响了另一部分建筑体的风格。

法国达飞海运集团总部塔楼（CMA CGM HEADQUARTERS TOWER）

法国，马赛，2006—2011年
与帕特里克·舒马赫合作

马赛是法国的第二大城市，这座古老的港口城市拥有悠久的航海历史。我们抓住设计这座新塔楼的机会，为这座城市竖立了一座垂直的地标，使其与城市中的其他标志性建筑遥相呼应。按照标准的高楼设计草案，往往将统一底板折叠以最大限度地缩减建筑耗时和耗材。我们偏离了这样的设计草案，着重塑造建筑围护结构以及室内中庭或入口，并使之搭配塔楼的设计造型：上层为惯常的办公空间，较低层则布置为延展的水平设施。这座建筑矗立于米拉博区域，靠近采石场和航道，这里遍布着战后的中高层建筑。这座建筑拔地而起，缓缓上升构成了弯曲的弧线状造型。浮现于地基基准之上的诸多载体构成了塔楼的建筑体，它们沿着构造柱塑造出双幕墙系统。在街面上，多模式的交通枢纽为行人提供了便利的公共交通。

德国古根海姆美术馆（DEUTSCHE GUGGENHEIM）

德国，柏林，2005年
与帕特里克·舒马赫合作

我们为这座美术馆所设计的隔板是镂空的实体造型，以便游客们能够穿梭于镂空的空间之中。根据所选展品的体积和数量，我们设定了椭圆造型的空间关系。随着游客逐渐深入展馆，椭圆造型也越来越富有动态性。当他们进入正厅之后，椭圆状造型变为了坚实的实体。在这里，它们化作夸张的壳状的水平与垂直的结构，向着玻璃顶棚延伸而去。

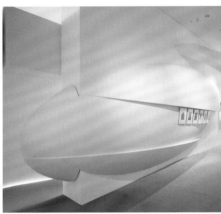

城市星云（URBAN NEBULA）

英国，伦敦，伦敦设计节，2007年
与帕特里克·舒马赫合作

我们设计这项装置的目的是为了探索混凝土的潜力。在运用传统的预先浇筑技术的基础上，我们综合了数控机压成型技术，利用混凝土塑造出重复、流动的造型。我们通常认为混凝土是一种惰性材料，然而，这项设计中的每一个部件均具有流动性，从而挑战了我们原有的概念。星云中有明亮的部分也有暗淡的部分，在这项设计中，通过排列这项设计中的每一个构成部件，我们将星云的这一特点展现为六边形和三角形的孔洞。

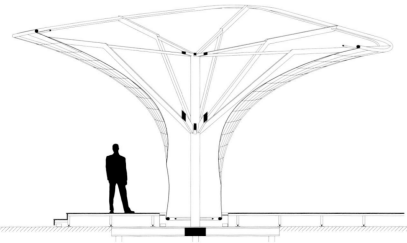

我们为蛇形画廊一年一度的夏季派对设计了这款造型的装置。三处造型一致的"阳伞"构造围绕着中心点组合出开放的空间。从小型的铰接底座缓缓上升,每一座阳伞构造皆向上延展为大型的悬浮状钻石造型。我们从花瓣等复杂的自然形状中获得了灵感,伞状结构相互重叠,在不碰触彼此的前提下相互交织。这座凉亭位于开放空地的低处,它婷婷而立,游客可以从任何一个方向进入亭子内部。白天,它可以为游客遮光;而在夜晚,亭子内部便能散发出柔和的光芒。

利拉亭(LILAS)
英国,伦敦,蛇形画廊,2007年
与帕特里克·舒马赫合作

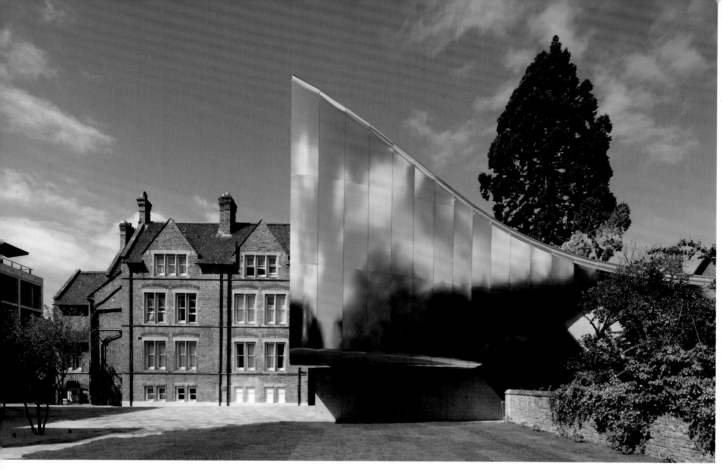

圣安东尼学院的中东研究中心收藏了牛津大学的现代中东重要藏品，这座世界级档案馆中藏有珍贵的私人文件和历史照片。这座投资公司大厦额外提供了1127平方米的底层空间，从而将中东研究中心图书馆与档案馆的扩建空间提升了一倍。与此同时，在这座大厦的地下部分，还建有一座可容纳117人的阶梯教室，通过地下换热器实现教室的通风散热。在图书馆档案室的下方，也装有相似的换热器，这样就能适当地控制环境，以保存研究中心的珍贵藏品。

我们所设计的这座投资公司大厦位于圣安东尼学院错综复杂的场地之上，与已存的国家保护级建筑以及树木交织在一起，但它们同时又保持着各自的独立性。图书馆阅览室的西面墙壁为曲面造型，以适应一棵百年红杉的枝干和它那繁茂的根系。我们在基础底板下安装了排水系统，以确保这棵红杉树水分的供给。档案馆阅览室和图书管理员办公室均朝向东方，它们拔地而起仿佛要超越立于一旁野兽派的希尔达·贝斯建筑（Brutalist Hilda Besse Building），其高度达到了这座建于1970年的建筑结构的屋檐。

投资公司大厦（INVESTCORP BUILDING）
英国，牛津，2006—2015年

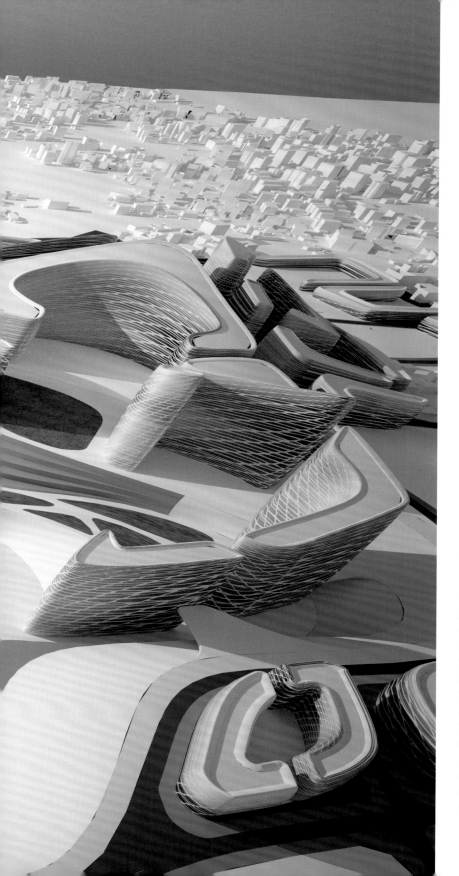

卡塔尔·彭迪克总体规划（KARTAL-PENDIK MASTERPLAN）

土耳其，伊斯坦布尔，2006年至今
与帕特里克·舒马赫合作

我们为卡塔尔·彭迪克设计了这项整体规划，目的是鼓励伊斯坦布尔这座城市向着多中心化的模式发展，以抵抗欧洲崇尚单一中心模式的偏见。这项总体规划将会带动经济的发展，为相邻的规划区域地带带来经济连锁效应。城市分区可以容纳100 000人，他们可以在这里生活和工作，促进这一地区的蓬勃发展。为了实现这样的效果，我们将街道设计为棋盘式布局，其中建有三处密集的中心群集区，以形成生动的天际线，将沿海和海洋特有的景色发挥至极致。一条主干道将成为这里的主动脉——在整个规划中发挥"脊梁骨"的作用——从而贯连北部的新地铁站和南部的铁路公路枢纽。北部和南部都将建立中心商业区，北部的采石场区域将发展为主要的娱乐区，娱乐设施将沿湖而建，周围将配置配套景观，而南部的滨海区将被发展为文化区。

在我们为迪拜湾的一座岛屿所设计的这片地标性发展规划中，包含一座可容纳2500人的歌剧院、一座可容纳800人的剧场、一座面积为5000平方米的艺术展览馆、一座艺术表演学校和一座六星级酒店。我们将所有这些设施囊括于一片建筑结构之中，并将其塑造成类似附近的沙丘的形状。两座中心尖顶的造型分别为歌剧院和剧场，尖塔造型向下蔓延与地面连接在一起，建筑轮廓凹陷处被塑造成了观众进入建筑的入口。主门厅的空间高度高出地面一层，如同徐缓的多重风景，这里被用作礼堂和艺术画廊。其上建有多间礼堂，在空间中营造出令人耳目一新的风景。

流动的造型包围着礼堂，这些造型看起来仿佛是从主壳状造型的下侧缓缓立起的。然而，位于内部的这座壳状造型并没有与主壳状造型相连，这两座壳状造型的外壳消失于彼此之间的缝隙之中。

迪拜歌剧院（DUBAI OPERA HOUSE）
阿拉伯联合酋长国，迪拜，2006年
与帕特里克·舒马赫合作

伊芙琳·格蕾丝学院（EVELYN GRACE ACADEMY）

英国，伦敦，布里克斯顿，2006—2010年
与帕特里克·舒马赫合作

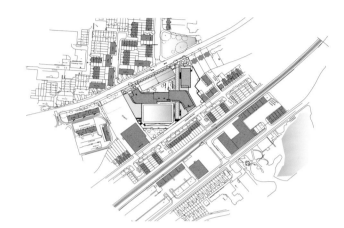

伦敦南部的这一片区域古老而活跃，这项设计则丰富了这里的教育多样性。这片区域原本主要为居民住宅区，该设计将这里的建筑环境融入了该社区的城市更新进程之中。这片建筑区域的关键部分位于两条主干道之间，从而赋予了该建筑以连贯的造型，使其具备很强的城市特色与个性，在当地以及相邻的区域中具有很高的辨识度。

该学院具备良好的学习环境，为学生们提供了充足的空间以及活跃的气氛，为教师们提供了积极的教学空间。为了贯彻"校中校"的教育原则，我们为多功能空间设计了独具特色的自然样式，并分别赋予四座规模较小的学校以独特的内在特色以及外在便于区分的特点。自然的灯光、通风设备和低调但耐久的纹理为这些空间营造出宽松的环境。在共享空间中——为所有学院所共享——设有多个汇聚点，它们将遍布全校的教学设施贯连在一起，鼓励着师生们进行社交活动。

为了营造出活跃的交流环境，基于多种交织的功能，我们将外部共享空间分层，以营造多层次的社交与教学空间。

东大门设计广场(DONGDAEMUN DESIGN PLAZA)

韩国,首尔,2007—2014年
与帕特里克·舒马赫合作

东大门设计广场一直被人们看作是一处文化中心。它所在的位置是首尔最繁华、最古老的区域之一,其内部建有一座设计博物馆、一座图书馆、诸多教育设施以及一座园景公园。各个年龄段的人都喜欢来这里——在这里他们可以交流思想,可以探索新的科技和媒体信息——在这里上演着瞬息万变的各种展览盛会,从而丰富了这座城市的文化生命力。考古所发掘的古老城墙和文化艺术品构成了这片广场不可分割的一部分。

一片稠密的城市中蔓生出的绿洲一样的公园一直延伸至建筑的屋顶处。公园平面上的空隙以及折射的光使游客们能够瞥见建筑内部的设计创新世界。诸如分层等传统的韩国花园设计元素被贯之以现代风情,因而避免了单一的特色占据整片视线。我们的设计将广场与公园天衣无缝地融合于延展的流动空间中,从而模糊了建筑与自然之间的界限。

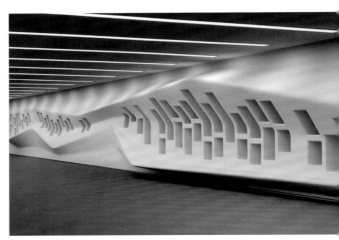

东大门设计广场

里吉乌姆海滨博物馆（REGIUM WATERFRONT）
意大利，雷焦卡拉布里亚，2007年至今
与帕特里克·舒马赫合作

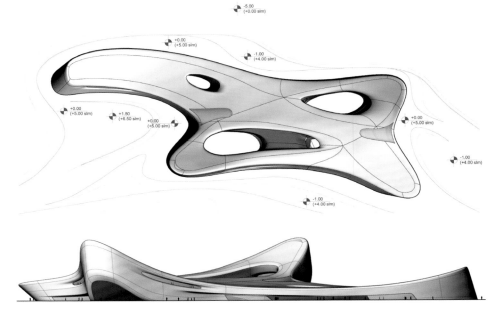

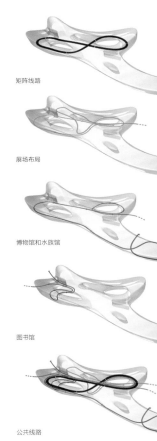

矩阵线路

展场布局

博物馆和水族馆

图书馆

公共线路

  地中海历史博物馆的多功能建筑位于西西里岛和意大利大陆之间的海峡处。在设计这座建筑时，我们的灵感来源于海星的造型，我们利用这项设计推进了对有机体形状的研究。极其对称的形状将不同的建筑部分以及设施相互贯连，自然系统中流动的造型也极具动态性和开放性。

  这座博物馆将包含展览空间、存储设备、一座档案馆、一座水族馆和一座图书馆。这座多功能建筑由三重建筑结构构成，以半包围的样式环绕着广场，将海滨区域引入建筑之中。

这座建筑位于墨尔本中心商业区西部的边缘，这里正发生着日新月异的变化。这座高楼应被设计为一座多功能的综合建筑，于是我们借助灵感将其整体结构分割为诸多规模较小的花瓶状造型，而它们则层层重叠为整体的建筑。在花瓶造型相接的地方，我们设计了公共空间，以便公众和市民以一种新的方式感受这座城市。

底层设有多家餐馆和一座新的公共展厅，最底部花瓶造型的顶端设有办公室和额外就餐区，人们可以从这里直接抵达位于第九层的公共露台。在这一公共楼层之上的花瓶造型内，设有各种各样的居住单元。

每一个花瓶造型都缓缓向下缩减体积形成锥形造型，从而在花瓶造型的底部营造出多余的露天空间。面向科林斯大街的一面设有优雅的廊柱，这些雕刻的弧形柱子支撑着这座建筑独特的立面系统。

科林斯大街582—606号（582-606 COLLINS STREET）
澳大利亚，墨尔本，2015年至今
与帕特里克·舒马赫合作

梅斯纳尔山博物馆 (MESSNER MOUNTAIN MUSEUM CORONES)

意大利，南蒂罗尔，2013—2015年
与帕特里克·舒马赫合作

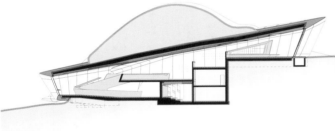

　　著名的登山家莱茵霍尔德·梅斯纳尔（Reinhold Messner）曾探索登山运动的传统、历史和规则，这是他所建的第六座也是最后一座梅斯纳尔山博物馆。在周边风景中尖锐的岩石以及寒冰碎片的启发下，我们利用就地铸造技术设计了混凝土天篷，使其从地面缓缓升起保护着博物馆的入口、观景窗和露台。外部嵌板由浅色的玻璃纤维加固混凝土构成，它们折入博物馆内部与深色的内嵌板相接。外部的嵌板映衬着科斯特地形所独有的色泽，内部的嵌板则散发着地表之下无烟煤般的黑色。

　　若干楼梯宛如山间瀑布一般从博物馆中伸出，将三层展览空间贯连在一起。自然光线穿过宽大的玻璃深深照入博物馆之内，吸引着博物馆内的游客走向全景窗边，走上最底层的观景露台。游客们可以穿过走廊走上悬吊式露台，这片露台从山体一侧向外伸出6米，使游客能够伸入阿尔卑斯山内部，以240度的视角欣赏山中的景色。

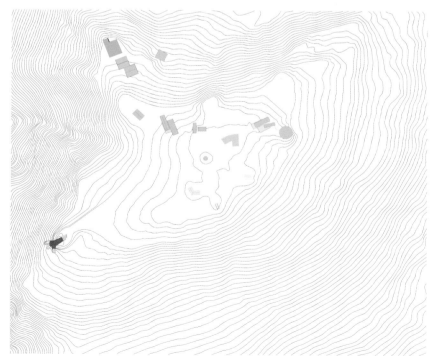

丽敦公寓（D'LEEDON）

新加坡，2007—2014年
与帕特里克·舒马赫合作

丽敦公寓位于新加坡第10区的中心地带，这片综合建筑包含7座居民高楼、12座半独立式别墅和配有娱乐设施的综合景观。高楼由上到下越来越细，以便在地面构造出私家花园。建筑被设计为独特的花瓣状造型，以便在房间的三面安窗，为住户提供最高质量的生活体验。厨房、厕所、客厅和卧室中均装有通风设备。高楼顶端化为手指状的造型，它们高度不一，模糊了建筑与天空之间的界限。

阿卜杜拉国王金融区地铁站 (KING ABDULLAH FINANCIAL DISTRICT METRO STATION)

沙特阿拉伯,利雅得,2012年至今
与帕特里克·舒马赫合作

阿卜杜拉国王金融区地铁站将成为新利雅得地铁网1号线的主要转换站,同样也将成为4号线(乘客可以乘此线前往机场)和6号线的终点站。人们也可以从这座地铁站出发,经过天桥,抵达单轨铁路站。共4层的公共楼层中建有6座站台,其地下两层还建有停车场,从而使这座地铁站与经济区的城市环境融为一体,在回应多模式交通中心的功能需求的同时,也顾及该区域的发展远景。

根据阿卜杜拉国王金融区总体规划的设想,这片区域将建设由公路、天桥和地铁构成的交通网。在我们的这项设计中,将地铁站置于总体规划交通网的中心。我们制作了该场地的周边互通地图和交通路线图,以便清晰地规划出建筑内行人路线,使建筑内部循环达到最优化效果以避免拥堵。最终的设计造型为三维格架造型,该造型由一排正弦波结构构成,正弦波结构如同脊柱一样支撑着建筑的循环系统,并延伸至建筑的外围。

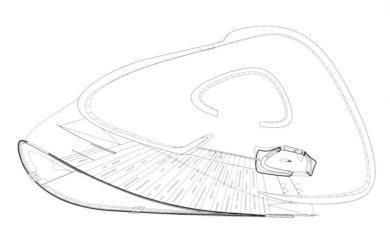

香奈儿流动艺术馆（MOBILE ART PAVILION FOR CHANEL）

香港，东京，纽约，巴黎，2007—2008年
与帕特里克·舒马赫合作

这座艺术馆的造型沿袭了香奈儿的典型作品形象，该品牌一贯注重在优雅连贯的整体中以精致的细节构成流畅的层次感。我们最终所设计的造型符合该品牌最初的灵感，无论在整体结构还是具体细节方面，该造型都是时髦别致、功能实用又多样的。这个造型的设计灵感来源于大自然中的螺旋状，我们运用参数化设计和周围的环境，塑造出各种各样的展览空间，并在造型中央设计了一座65平方米的庭院以供游客游览。当游客穿梭于艺术馆内部空间的时候，这样的设计使他们能够看见彼此的行为，从而引导他们体验这场艺术之旅。

壳状造型由诸多逐渐缩小的拱形零件构成，其可拆卸零件的宽度都不超过2.25米，这样非常便于运输。外墙上的分隔缝构成了这座建筑的显著造型特征，它为整座建筑营造了空间节奏感。能量般的线条在亭状建筑内部交织，持续地定义着每一处空间的性质，引导着人流在展馆中的动向。我们所面对的挑战在于，如何将知识和形体层面的设计转换至感官层面——为了彰显标志性流行品牌的魅力，我们将实验前所未有、身临其境的环境，使这件艺术品在世间经历持续不断的重塑。

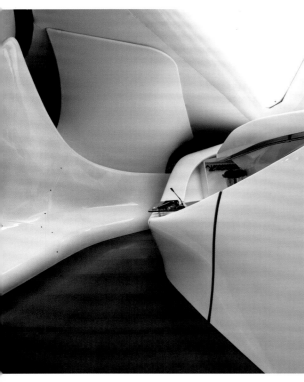

香奈儿流动艺术馆

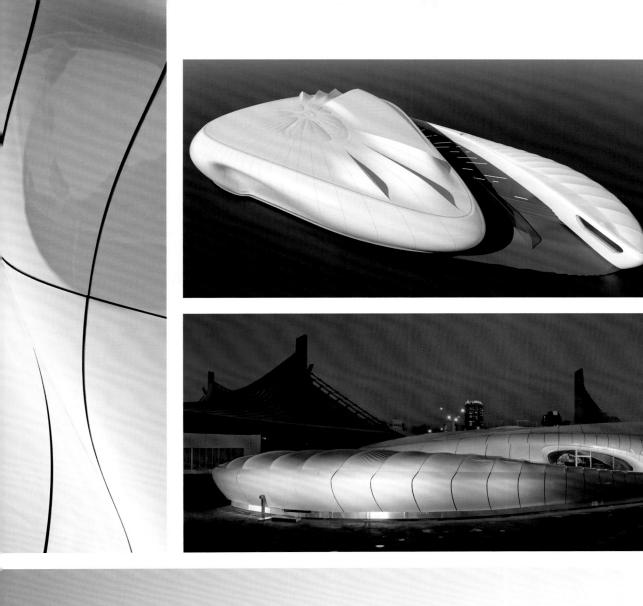

阿利耶夫文化中心（HEYDAR ALIYEV CENTRE）

阿塞拜疆，巴库，2007—2012年

与帕特里克·舒马赫合作

我们满怀雄心地设计了这个作品，希望它能够在这座城市的市民精神生活中发挥整体性的作用。我们为这座建筑设计了流动的造型，使其浮现于周围风景那层叠的地貌之间，并将这座文化中心的各项功能汇聚于一体。各项功能以及各个入口均被置于单一延展的墙面褶皱之中。流线型的造型不仅容纳了各式各样的空间，而且赋予每一处空间以独特性和私密性。随着墙壁上的褶皱向内延伸，墙体逐渐化作内部构景的一部分。

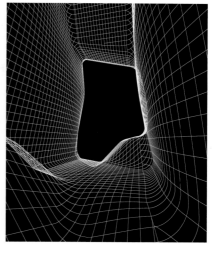

这座建筑坐落于迪拜商务港总体规划区,这座奥姆尼亚公司投资建设的旗舰高楼由两部分建筑体构成,这两部分建筑体则由延展的低洼裙房结构相连接。裙房结构位于两座建筑体的底部,将二者组合为统一的整体:一个立方体造型拔地而起,其内部为自由样式的镂空造型。立方体结构由传统的平板垂直层叠而成,其中心处设有孔洞,以便充分利用建筑立面附近的区域。侵蚀状中空造型之上覆盖了双层着色玻璃,其本身也是建筑体的一部分,一直向上延伸至该建筑的边缘。中空造型中的起伏处为游客们提供了娱乐和休息的场所。底层被塑造为露天的场地,以吸引游客进入这两座各自独立的大厅。玻璃立面上镶有反光的像是融化了一样的花纹,由矩形马赛克构成,从而使该建筑具备反光性和材料特性,以减少太阳光的辐射。在白天,立方体满富活力,中空体空无一物,而在夜晚,却营造出相反的视觉效果:黑色的立方体看起来失去了实体形状,而被灯光点亮的中空体即使在远处也清晰可见。

"奥普斯"大楼('OPUS' TOWER)
阿拉伯联合酋长国,迪拜,2007年至今
与帕特里克·舒马赫合作

赛马会创新楼（JOCKEY CLUB INNOVATION TOWER）

中国，香港，2007—2014年
与帕特里克·舒马赫合作

赛马会创新楼隶属于香港理工大学的设计学院和赛马会社会创新设计院。这座建筑高达15层，可容纳超过1800名的学生和工作人员，配备了用于设计教育以及创新的设施，其中包括工作室、实验室和车间、展览区、多功能教室、一座阶梯教室和一间公共休息室。

赛马会创新楼位于校园东北角一处狭窄的不规则场地，将塔楼裙房的一贯造型消解为更为流畅的构造。建筑内部与外部院落则营造出不规则的空间，它们彼此之间相互贯通和互动，随着学生、工作人员和游客在建筑中走动，建筑内部的玻璃窗以及空间营造出透明的连贯性。在设计学院中，这项设计将诸多程序贯连在一起，从而营造出多学科的环境，塑造了综合的研究文化，使各种贡献与创新融会贯通、齐头并进。

多米尼恩办公楼 (DOMINION OFFICE BUILDING)

俄罗斯,莫斯科,2012—2015年
与帕特里克·舒马赫合作

多米尼恩办公楼坐落于沙里库普得施普尼克夫斯卡亚大街(Sharikopodshipnikovskaya Street)之上,它位于莫斯科前工业区和居住区之中。随着创新区和信息技术区在这一地区的蓬勃发展,一系列的新建筑也在此地陆续建成,而这座多米尼恩办公楼便隶属于所建成的第一批建筑。这座建筑由诸多层叠板垂直堆积而成,每一层层叠板均与上一层相错开,它们以弧形元素连为一体。一座中央大厅贯通了所有的楼层,从而将自然光线引入建筑物的核心部位。每一层的阳台均指向大厅,与其外部外围结构相互呼应,而诸多楼梯则穿梭在中心空间中,形成了相互交织的通道。大厅底层的餐厅不仅将大厅与外部的台阶和主要街道连为一体,而且将咖啡零食区与阳台上的休息区连为一体。这座大厅在几层楼之间形成了共享的空间,从而引导着建筑内部公司员工的相互交流。办公空间被安置于标准直线型小开间中,从而为小型、拓展型以及大型公司提供了多种不同的可能性。

伊莱和伊迪萨艺术博物馆 (ELI & EDYTHE BROAD ART MUSEUM)

美国,密歇根州,东兰辛,2007—2012年
与帕特里克·舒马赫合作

  这座艺术博物馆位于密歇根州立大学,我们最初的设计理念是塑造一片"风景式地毯",从而将建筑区域松散的城市结构连为一体,并将其沿着不同的运动方向编织在一起,使其沿着建筑场地相互交织。周围的结构营造出了线性的走势,从而为这片地毯提供了场地基础,并将其延伸至建筑体本身之上。综合建筑的每一个侧面均面向不同的方向,将建筑外表塑造为棱角分明的主体。这座建筑以夸张的姿势向西方倾斜,形成了凸起的顶部造型,而高达12米的前端面则俯瞰着广场,进而以线性的变化转向东方,面向雕塑园,并与风景融为一体。玻璃和不锈钢外部表层遍布着空隙,呈现出不断变换的表面造型,从而激活了整座建筑。变换的表面也过滤和引导着每一座展馆中的日光。

淡江大桥（DANJIANG BRIDGE）

中国，台北，2015年至今
与帕特里克·舒马赫合作

  淡江大桥坐落于流经台北的淡水河河口，它是台北向北部发展建设的基础设施中不可分割的一部分。这座大桥将加强相邻设施之间的连贯性，而通过连通数条主要道路，它将减轻当地市中心的道路交通压力。这座斜拉桥设计造型只使用了一根混凝土塔柱，以支撑长达920米的钢筋道路、轨道和人行桥面，将对天际线轮廓的影响降为最低。它将成为全世界最长的单塔不对称斜拉桥。

阿布扎比表演艺术中心
(ABU DHABI PERFORMING ARTS CENTRE)

阿拉伯联合酋长国,阿布扎比,2008年至今
与帕特里克·舒马赫合作

　　这座艺术中心的造型源自自然界的组织系统,它从属于萨迪亚特岛总体文化机构。我们利用生长模拟过程营造了一系列的基本形状,随后在方案图的协助下,将其附加为一系列循环往复的结构。于是,这些与生物相类似的最初构成元素(枝叶、茎干、果实和叶子)从抽象的图案被转换为建筑造型。这一文化区的核心轴是一条行人走廊,它从谢赫扎耶德博物馆一直延伸至海边。这条轴线与海滨长廊相互呼应,共同营造出树枝状的造型,而岛屿则分布在独立的空间中,以容纳五座主厅。

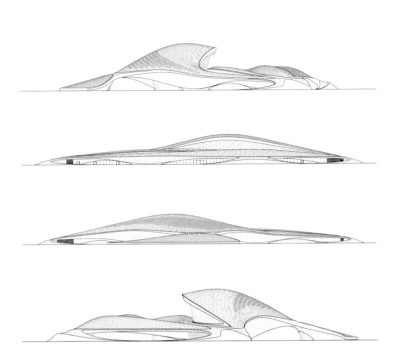

毕阿总部（BEE'AH HEAD QUARTERS）

阿拉伯联合酋长国，沙迦，2014年至今
与帕特里克·舒马赫合作

毕阿是位于中东的一家全方位综合性环境保护和垃圾处理公司。这座新的总部建筑属于这家公司尚在修建工程的一部分。通过提供基础设施、工具和支持，公司旨在以此改变人们的态度与行为，从而实现其环保的目的。

这座新的总部临近公司的垃圾处理中心，它由诸多纵横交错的沙丘状造型构成，旨在顺应盛行的夏马风的走势，并为其内部空间提供尽可能充足的光线和尽可能广阔的视野。两座中心沙丘造型中包含了办公室、游客中心和高管部门。一座中心院落将这两座沙丘造型连为一体，从而在建筑物内部营造出"一片绿洲"以供人们使用。

在设计这座建筑时，我们充分考虑了每一层面的环保问题。建筑供电百分之百来源于可再生能源。建筑表面的选材则考虑其反射阳光光线的能力。建筑物的部分表面及其结构均符合标准正交维度，建筑物的重要部分则由再生材料构成。新总部建筑系临近十号工作室（伦敦基础环境设计顾问十号工作室），这样便可将这两座建筑用于冷却和饮用水的能源降为最低。

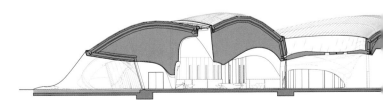

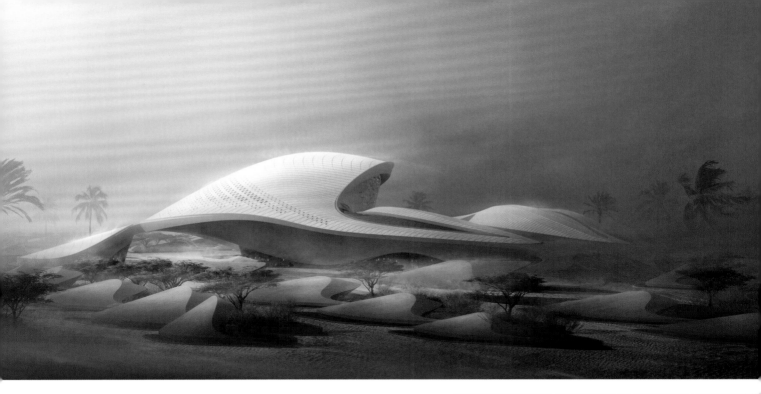
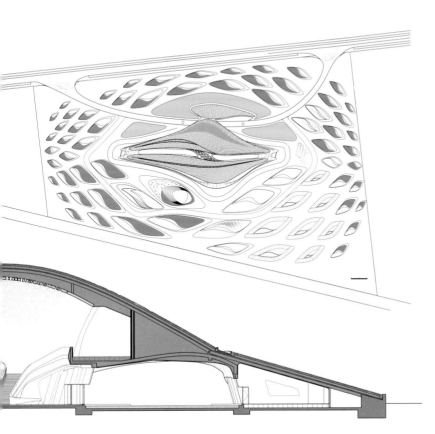

港口之家（PORT HOUSE）

比利时，安特卫普，2009—2016年
与帕特里克·舒马赫合作

　　这项为安特卫普港务局所设计的新总部包含两座建筑实体：一座原有的消防站和一座新的"悬浮"于其上的水晶建筑。二者一同组成了新的标志性建筑，俯瞰着整座城市和港口，给人留下难忘的印象。

　　建筑物正面的三角形玻璃面板构造了新的延伸部分，使其以不对称的方式悬浮于消防站的中心院落之上，建筑的侧面在白天倒映着天空中变幻的光晕，在夜晚则将整座建筑变换为光芒四射的水晶体，象征着安特卫普历史上兴盛的钻石产业。建筑南端的面板是平坦的，却在渐往北去的过程中变得越来越尖锐，于是建筑一面平坦，另一面则呈波纹状。

　　旧消防站的中心院落之上被加盖了玻璃的屋顶，从而将其转换为新港口之家的主接待区域。游客们能够从中央大厅抵达历史公共阅览室与博物馆，而后者则位于废弃的消防车大厅，我们对其进行了细致的修复和保护。一座外部大桥将原建筑与延伸部分贯连在一起，从而为我们提供了全景视线。

图书馆与学习中心（LIBRARY & LEARNING CENTRE）

奥地利，维也纳，经济大学，2008—2013年
与帕特里克·舒马赫合作

这座新图书馆位于维也纳经济大学的校园中央，它拔地而起形成了一座多边形块状建筑。这项设计采用了立方体造型，垂直立面和倾斜立面相结合。建筑外部的直边线在向内延伸的过程中分开，化作流畅的曲线以塑造出造型自由的内部峡谷，从而塑造出位于建筑中央的公共广场。宽敞的斜坡和楼梯从入口处向上延伸穿过了中央图书馆，呈漏斗状延伸，向上穿过了建筑的6层空间。建筑最上面的两层为阅览室和学生研习区域，在这里可以俯瞰普拉特公园的景色。这座新综合建筑的其他配套设施则置于另一单幢建筑体之中，这座建筑同样也一分为二，化作两排丝带状的建筑造型，彼此环绕着圈出中央流光溢彩的集会空间。

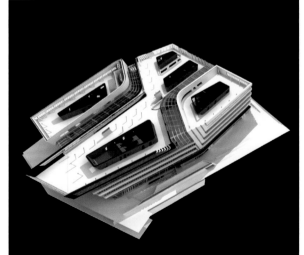

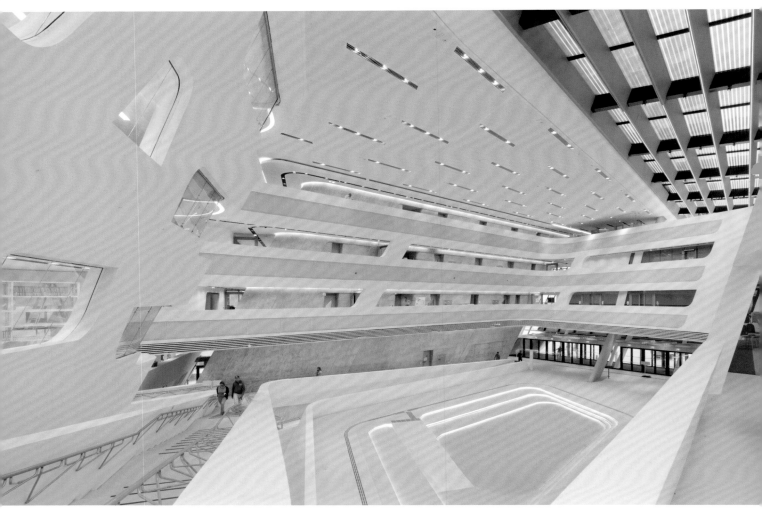

图书馆与学习中心

伯纳姆馆(BURNHAM PAVILION)

美国,伊利诺伊州,芝加哥,2009年
与帕特里克·舒马赫合作

芝加哥以尖端建筑与工程而著称,我们所设计的这座伯纳姆馆恰恰凸显了芝加哥的这一特色。在设计这座建筑造型时,我们将新的造型概念融入以往大胆的城市规划之中。这座新建筑由错综复杂的弯曲铝合金结构构成,其每一构成部分均被塑形并焊接,从而营造出独特的曲线造型。金属构架的内部和外部均被紧紧地包裹了一层布质材料表皮,以塑造出流畅的造型,如同屏幕一般反射着视频设备的投影。这项设计能够最大限度地重复使用可再生材料,当它在百年庆祝中完成使命之后,可被重置于其他场地之上,以便在未来重复使用。

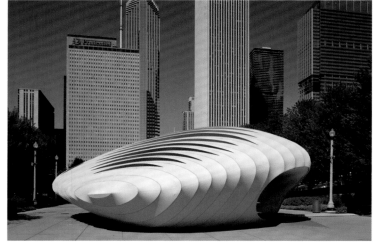

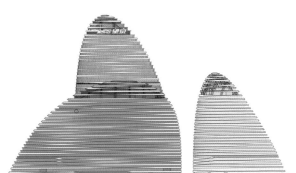
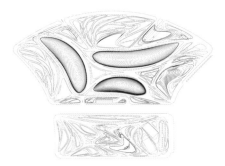

望京SOHO（WANGJING SOHO）

中国，北京，2009—2014年
与帕特里克·舒马赫合作

这座建筑距离北京首都机场与市中心的距离基本相等。该办公和商铺综合建筑由三座弧形的塔楼构成，高度从118米至200米。这三座塔楼被设计为相互交织的"山脉"，它们将建筑与风景融为一体，以创造出新的公共空间，从而为当地的市民提供了聚集的空间。其南部为一座公园，北部为风景式花园。每一座塔楼入口处宽敞的前厅均面向市区，它们向外延伸直至中心商业街和塔楼之间的广场。随着视角的变化，塔楼并置的流动造型也处于不断的变化之中。于是，在某些角度观看的时候，它们是独立的建筑，而从其他角度观看的时候，它们是一座相互贯连的集群建筑。

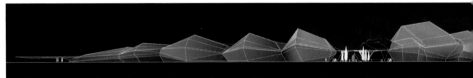

在设计这座研究中心时,我们充分考虑了环保因素。与此同时,我们也致力于设计出有机造型,使其超越简单的技术策略,在保留这项设计的视觉特色的同时,我们容许它在造型风格方面尽情延展和变形。由此,我们创造出一座水晶造型的多孔结构。像晶体般生物细胞结构一样的建筑从沙漠中拔地而起,适应着当地的气候环境和其内部的功能限制。

这项设计的所有元素均遵循一致的空间和结构策略。每一座建筑均被分配了各自的功能,它们可以在必要的时候随机应变。我们在建筑外部塑造了坚硬而结实的外壳,而建筑内部则建有多座荫蔽院落,将一定量的日光带入空间内部之中。建筑中建有诸多分层和缓冲地带,将炫目耀眼的日光温柔地过渡至凉爽净化的建筑内部。

阿卜杜拉国王石油研究中心(KING ABDULLAH PETROLEUM STUDIES & RESEARCH CENTER)

沙特阿拉伯,利雅得,2009—2016年
与帕特里克·舒马赫合作

银河SOHO (GALAXY SOHO)

中国，北京，2009—2012年
与帕特里克·舒马赫合作

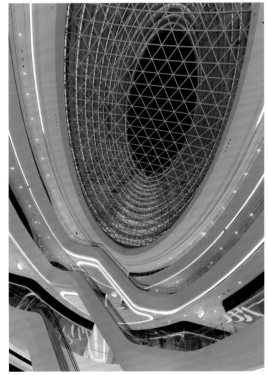

我们为SOHO在中国设计的是一座集办公、零售和娱乐为一体的综合建筑，其灵感来源于北京宏大的城市规模。这个项目由五座连贯而流动的建筑体构成，它们各自独立，由延展的桥梁连接在一起。它们在多个方向上与彼此呼应，营造出全景式建筑效果。建筑体上既没有棱角也没有突兀的过渡，以避免打破造型的流畅性。

这项设计的内庭致敬了中国传统建筑的院落，从而在建筑内面塑造出延展的开放空间。各建筑体在此处相互交融，从而使每一座建筑均与彼此呼应，共同塑造出流畅的走势。设计内部变换的建筑走势营造出强烈的融合感和韵律感。随着游客们逐渐走入建筑内部，他们会发现其内部空间遵循了一致的造型逻辑——持续的曲线性。

建筑空间的底三层用于零售业和娱乐业，其上的楼层则为群集的商业提供了工作空间。建筑的顶层修建了酒吧、餐馆和咖啡厅，在这里可以俯瞰这座城市中极其繁华的一条街道。一系列配套设施将不同的功能集于一体，这在都市中是司空见惯的现象，由此，这座银河SOHO成为北京的一个主要都市地标。

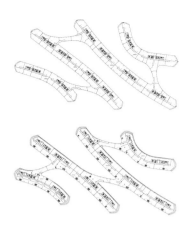

这座办公与零售综合建筑坐落于虹桥商业区的凌空经济园中,由此可以通往高铁和地铁系统,方便往来于机场与市中心。

这项设计概念着重强调四面平行板低调的简洁性。简单的模块化系统建于8.4米的格栅之上,其建筑深度达18米,设计中尽显便利性与灵活性。四座主建筑构成了环绕的建筑造型,蜿蜒的零售商铺将这些细长的元素贯连在一起,一条延展的金属丝带则将它们环绕为恢宏的走势,其上还覆盖着绿色的屋顶。

一片"挤压空间"与以上所提及的简单构成元素形成了鲜明的对比,设置在中间区域,设定为内部和谐温馨的庭院,在主建筑之间桥梁的作用下,它们贯连在一起,形成了独特的造型。由此,在欣赏动态的内部和外部公共空间的时候,可以获得浑然一体的空间体验。

凌空SOHO(SKY SOHO)
中国,上海,2010—2014年
与帕特里克·舒马赫合作

凌空SOHO

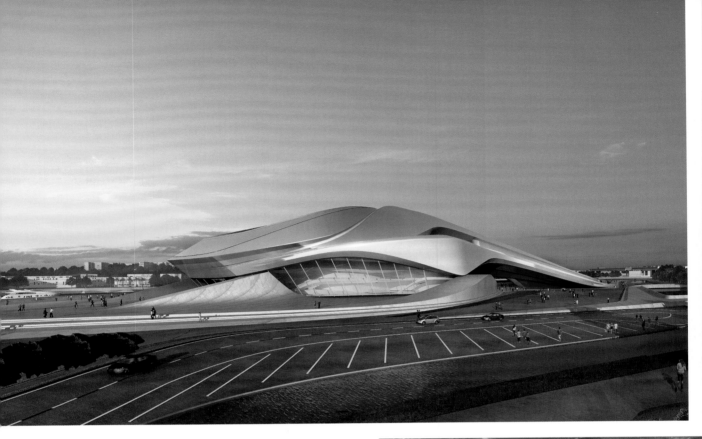

拉巴特大剧院（GRAND THEATRE DE RABAT）

摩洛哥，拉巴特，2010年至今
与帕特里克·舒马赫合作

这项设计的场地位于临近的两座城市——拉巴特与塞拉之间，其西北部为旧欧代亚城堡，东北部为机场，西南部为皇家宫殿，东部为规划中的巴林发展区，而布赖格赖格河则定义了这一建筑所在园区的地势。这座建筑造型流畅，它拔地而起向着天空延展，以营造出包围式的效果，随后该建筑造型顺势覆盖了其下的两座礼堂，最后以拱形结构与地面相接，融入周围的风景之中。

在声学原理和配套设施的背后，这类造型还顾及了建筑的经济节约性。建筑的密实度足以使露天圆形剧场免受日照以及来自于穆莱哈桑大桥的噪声的影响，并排的三座剧院可以共享配套设施，从而减少了所需的空间。雕刻造型提供了天衣无缝的空间体验，它们以流动的形式遍布于主厅之中，塑造了气势恢宏的楼梯，为游客们提供了直观的视觉引导。

毕尔巴鄂比斯开储蓄银行总部 (BBK HEADQUARTERS)

西班牙，毕尔巴鄂，2010年至今
与帕特里克·舒马赫合作

为毕尔巴鄂比斯开储蓄银行（Bilbao Bizkaia Kutxa）所设计的这座新总部位于佐佐塔雷半岛（Zorrotzaurre peninsula）的东南角。由于该地在建筑界享有盛名，从而需要以一座建筑着重标注此地的重要性。我们提议构建一座雕塑造型，由两座星状的设计造型构成。在塔楼的顶部和底部，两座星状造型以45度角与彼此相交，由此一座星状造型的外尖变成了另一座星状造型的内角。所构造的造型随着视角的转换而不断发生变化。

整体结构架进一步定义了这座建筑的造型，它既塑造了基础结构，又为立面附属结构提供了支撑系统，从而使建筑内部不必再使用支柱。在塔楼顶部设有全景式观景楼层，构成了综合办公室的一部分，塔楼底部则与花瓣造型的裙房融为一体，其中包括入口大厅、一家银行分行、一座体育馆和一间慈善活动办公室，从而使塔楼能够直接与地面互动。

OTM住宅大楼 (ONE THOUSAND MUSEUM)

美国，佛罗里达州，迈阿密，2012年至今
与帕特里克·舒马赫合作

这座极其奢华的住宅大楼高达66层。置身于建筑中，你可以望见从比斯坎湾到迈阿密海滩的风景。一座混凝土骨架结构沿着大楼的边缘向上攀升，构成了网状的流线造型，由此建筑内部的底层板上不必再立任何基柱。在大楼的底部，骨架结构伸展的流线在夹角处相交，构成了刚性的管状造型，以抵御迈阿密要求较高的风荷载。建筑底层中包括商业空间和停车场，其上建有公寓，最高的两层住宅为复式阁楼。在建筑的顶层中建有水上运动中心、休闲区和自由活动空间。

物品、家具和内部装饰

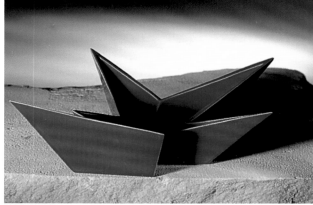

魏克森博格陶瓷
Waecthenberg Ceramics

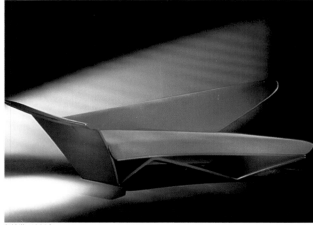

红沙发, 1988年
Red Sofa, 1988

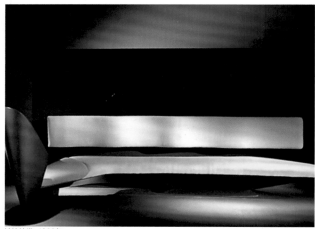

波浪沙发, 1988年
Wave Sofa, 1988

扭面灯, 1987年
Warped Plane Lamp, 1987

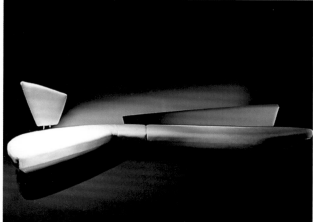

沃诗沙发, 1988年
Whoosh Sofa, 1988

早期家具和物品
多位制造者 1987—1990年

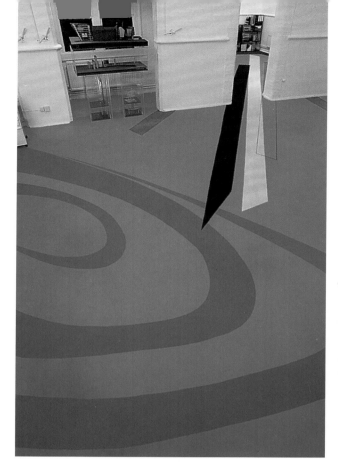

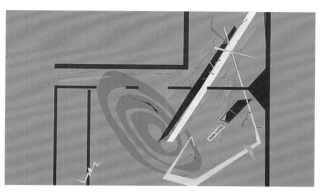

福维克满铺地毯，1990年
Vorwerk Wall-To-Wall Carpeting, 1990

致敬维奈·潘顿，1990年
Hommage à Verner Panton, 1990

茶与咖啡用具套装（TEA AND COFFEE SET）

萨瓦亚和莫洛尼（SAWAYA & MORONI），1995—1996年

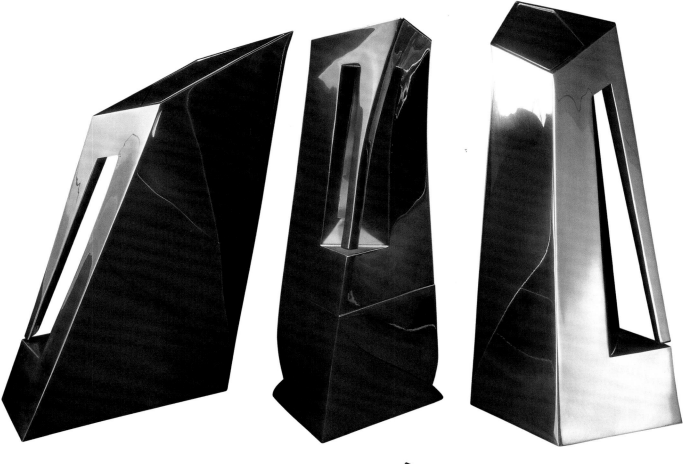

这件为萨瓦亚和莫洛尼所设计的桌上雕塑由四部分各自独立的元素构成：茶壶、咖啡壶、牛奶罐和糖碗。如同一副立体的组合七巧板，这四部分元素拼合在一起组成了一个完整的整体。当我们不使用这套用具的时候，这些不规则的物体便可以被拼合为一个完整的结构，我们可以将其视为单个的物体。当我们使用这套用具的时候，便可以将其拆成"四分五裂"的样子，让这些分散的物体发挥各自的功能。如此，这项设计体现了物体之间的互动，探索了几何体之间相互作用的新的可能性，研究了家用器具的动态雕塑造型。这项设计所提供的经验，同样也可以用于大型建筑。

这款茶与咖啡用具造型如同液态的金属，它试图重新定义日常用品的仪式性意义。通常情况下，这些用具的各部分是各自独立的。但在这项设计中，它们被组合成了一个可重构的整体。我们可以将这款用具"静置"为一件雕塑体，也可将其作为动态元素发挥其"功能"。这些组合构件静置在一个托盘之中，使用者可以按照多种方式将其组合，用具在使用时和静置时也分别呈现不同的造型。在形式上，这件雕塑造型的用具探索了垂直物体与水平物体之间的对比与组合。茶壶的有机形状宽大而平滑，咖啡壶则如同一座高塔于风景之间拔地而起。在功能层面，按照托盘所给出的模版，我们可以打开或旋转用具构件。每一个造型之上都有一处切口或者是一个零件，使得构件可以滑至一条不同的轴线上，由此一方面让使用者倒出液体，另一方面则塑造出完全不同的造型。于是，"饮茶时间"被赋予了新的意义：去解开这个雕塑谜题。使用者滑动、翻开并转动着构件，他们在茶盘中寻找着答案。

皮亚扎茶与咖啡用具（TEA AND COFFEE PIAZZA）
艾烈希（ALESSI），2003年
与帕特里克·舒马赫合作

Z-斯凯珀（Z-SCAPE）

萨瓦亚和莫洛尼，2000年

这款紧凑的集成设计将客厅家具组合在一起，其背后的造型理念来源于风景构造，它将不同的碎片构件组合为一个综合的整体。在塑造这11件构件的形状时，我们考虑了造型、功能以及环保层面的因素。然而，我们并没有使用先前所决定的模式，而是赋予了构件以一定的陌生性以及不确定性。如同智力拼图一般，这些构件被组合在一起，我们可以将其拆开，重组成不同的造型。比如，这件用于休息的"盒子"，便是流动的软硬交织空间中所拼合的一部分：软空间贴合地面，被塑造成了舒适的座椅；而硬空间顶部为纵向的平缓表面，被塑造为餐桌、架子、书桌和吧台。

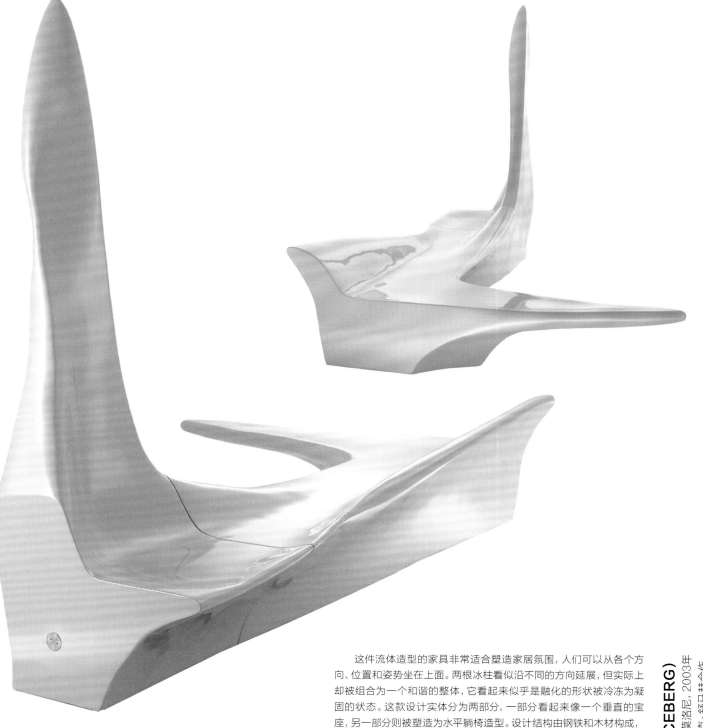

这件流体造型的家具非常适合塑造家居氛围，人们可以从各个方向、位置和姿势坐在上面。两根冰柱看似沿不同的方向延展，但实际上却被组合为一个和谐的整体，它看起来似乎是融化的形状被冷冻为凝固的状态。这款设计实体分为两部分，一部分看起来像一个垂直的宝座，另一部分则被塑造为水平躺椅造型。设计结构由钢铁和木材构成，其上覆盖着珍珠白色的汽车用漆，我们这样设计的目的是为了最大限度地体现人体工程学轮廓。

冰山（ICEBERG）
萨瓦亚和莫洛尼，2003年
与帕特里克·舒马赫合作

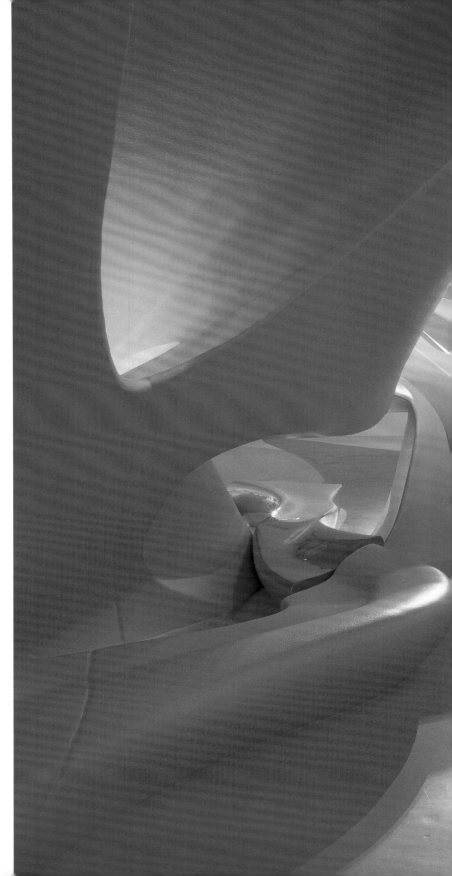

冰暴（ICE STORM）
奥地利应用艺术博物馆（ÖSTERREICHISCHES MUSEUM FÜR ANGEWANDTE KUNST），2003年
与帕特里克·舒马赫合作

这件家具宣扬了新的居住和休息环境，它综合了我们先前所设计的家具元素以及设备：Z-斯凯珀（第264页）、冰山（第265页）、蛇形造型和室内波浪，还包括冰流在内。诸多元素被汇集至动态的旋涡之内，两座特意设计的新硬沙发将它们合为一体。半抽象造型表面构成了一间公寓，仿佛是从蔓延的单一物体上镂刻出来的，其内遍布着褶皱、壁龛、洞穴和凸出物。通过"造型"技术，这项设计强调了复杂的弧度、无缝的衔接和平缓的过渡，由此将原本寻常的家具构件转换为统一的构造，从而构成了规模更大的结构。与整体造型不相称的元素（比如蛇形造型）便呈现为碎片状，被悬浮在场景的外围。游客们应邀居住在设计结构之内，他们会在此探索开放的美学，由此而重新思考我们业已接受的家居生活方式和行为概念。

贝卢板凳（BELU BENCH）

肯尼·沙赫特（KENNY SCHACHTER），2005年

与帕特里克·舒马赫合作

这件家具造型看起来像一个自成体系的单一物体，它可以适用于多重用途，可以被用作桌子、柜台、椅子或者容器，你甚至可以直接倚靠在上面。我们赋予了它复杂的动态几何造型，其周身散发着流动的特性。通过与人体直接接触，贝卢板凳可以创造出变化多端的周围环境。它不仅仅是一个用于展示的物品，还是一个动态的造型；由此，在空间上定义着周围的环境，同时也为我们提供了各种各样的功能。回顾最初的设计理念，虽然我们保留了那些独特的造型，却将其在三维空间中按比例缩小。这一系列的作品进一步提升了可能性，让我们直接介入家居环境。

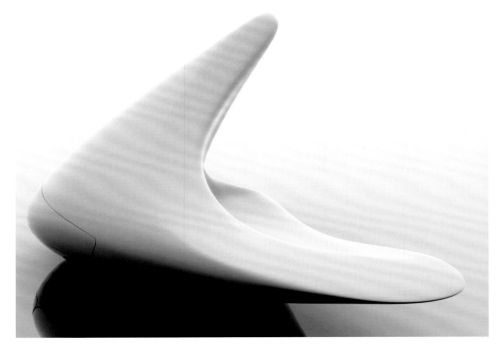

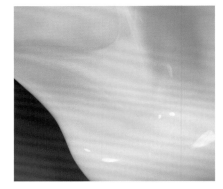

扎哈·哈迪德60号碗、70号碗和有机化合物（ZAHA HADID BOWLS 60, 70 AND METACRYLIC）

萨瓦迪亚和莫洛尼，2007年

与帕特里克·舒马赫合作

这个银碗具备流线的弧形造型，它回应着来自于其自身的分散能量，由此引导我们探索自然力，并与环境之间建立互动关系。虽然一眼望去，这只碗看起来浑然一体、自成体系，但实际上它整体的流动性却遵循着我们所研究的至高造型逻辑，从而进入持续转换与平稳过渡的体系之中。

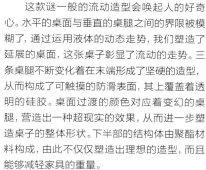

浅绿色桌子（AQUA TABLE）

艾斯塔子弟家具公司（ESTABLISHED & SONS），2005年

与帕特里克：舒马赫合作

这款谜一般的流动造型会唤起人的好奇心。水平的桌面与垂直的桌腿之间的界限被模糊了，通过运用液体的动态走势，我们塑造了延展的桌面，这张桌子彰显了流动的走势。三条桌腿不断变化着在末端形成了坚硬的造型，从而构成了可触摸的防滑表面，其上覆盖着透明的硅胶。桌面过渡的颜色对应着变幻的桌腿，营造出一种超现实的效果，从而进一步塑造桌子的整体形状。下半部的结构体由聚酯材料构成，由此不仅仅塑造出理想的造型，而且能够减轻家具的重量。

流动（FLOW）

塞拉伦加（SERRALUNGA），2006—2007年

与帕特里克：舒马赫合作

在搭配旋转模塑技术的前提下，通过复制三维造型技艺，我们利用这项设计进行了一场实验，以展现产品设计与塑造中的一项新类型：一件弯曲迂回的物品，其延展的表面上融合了笛卡尔式几何图形（Cartesian geometries）。

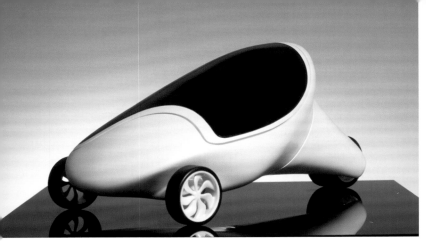
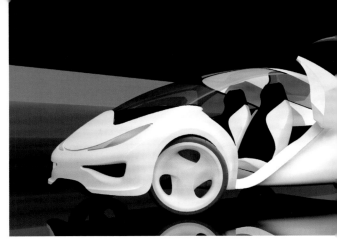

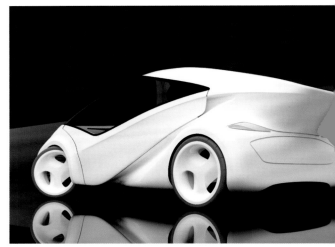

第一代与第二代Z-汽车（Z-CAR I AND II）

肯尼·沙赫特（KENNY SCHACHTER），2005—2008年
与帕特里克·舒马赫合作

第一代Z-汽车由氢燃料电池驱动，可以搭乘两人，其上装有三个轮子。它的升级版第二代Z-汽车装有四个轮子，可以搭乘四人。这两代汽车的流线造型综合了各种功能、静音操作以及气动性能。它们均为零排放汽车，通过可充电锂离子电池驱动。它们均配备有车轮内电动机，可节省空间、提高性能。在整体设计中，重量和空间被高度分配至所有的机械与电动配件之中。

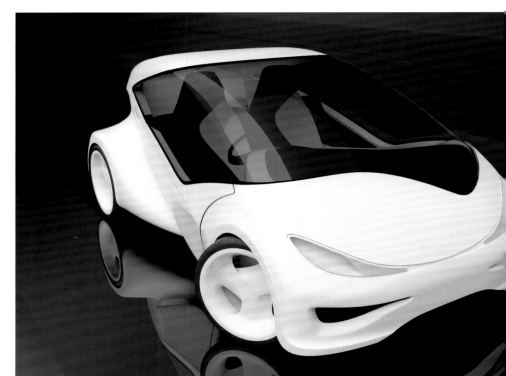

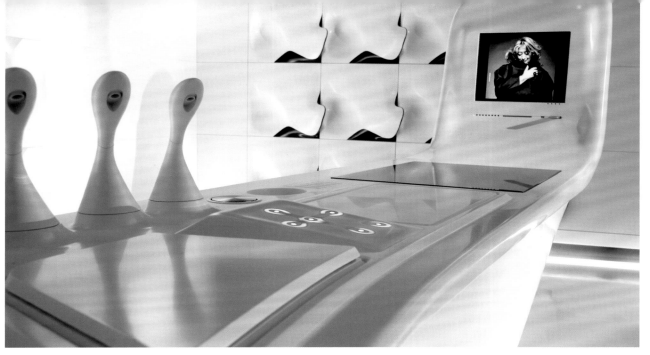

通过观察诸如冰体融化以及冰川移动等自然现象，我们设计了这一家居系统，从而将设计理念发挥至极致，以塑造出延展的人体工程学配件，使其天衣无缝地符合厨房的需求。这座组合设计中包含一片主设计，它被塑造成了悬臂构造，从水平的烹饪设备过渡至就餐设备进而过渡至垂直的数码表面；而次级设计则包含清洗池、洗碗机以及支架配件，其背景由波浪形的元素构成，它们以不同的方式循环、组合以形成复杂的纹理。镀层之所以选择可丽耐品牌，是因为它不仅具备可作热压成形的性质，而且还具备透明性和持久性。主设计充分体现了21世纪的电器特色：我们配备了多达2000盏LED灯，以展示各种各样的信息，而隐形的电热膜发挥了为食物保温的功效。

Z-岛案台（Z-ISLAND）

杜邦可丽耐（DUPONT CORIAN），2005—2006年

与帕特里克·舒马赫合作

无缝聚集（THE SEAMLESS COLLECTION）
艾斯塔子弟家具公司，2006年
与帕特里克·舒马赫合作

在这一套组合家具中，我们在家居范围内，进一步探索了设计界的无缝流畅性。近期三维设计软件以及制作工艺的进步为我们提供了动力，这套设计复杂的弧线形状反映了我们所做的人体工程学的细致研究，使我们思考家具与空间之间的平衡。每一件家具作为独立的物体均发挥着各自的作用，它们组合在一起又会形成感官的宇宙。在这个宇宙中，柔软与尖锐相遇，凸面与凹面交锋，就像是一片片的碎片被吸入了引力场之中。正如我们在Z-斯凯珀（第264页）、冰暴（第266页至第267页）以及浅绿色桌子（第269页）中所应用的那样，这项设计中的折叠、壁龛、凹陷以及凸起均遵循着连贯的构成逻辑，并采用了美国普尔塔酒店中的内部装修（第155页）。

沙丘组 (DUNE FORMATIONS)

大卫·吉尔画廊 (DAVID GILL GALLERIES), 2007年
与帕特里克·舒马赫合作

我们为威尼斯双年展所设计的这项作品有机地结合了诸多家具元素,其中包括架子、桌子、椅子和假树。这些家具的每一部分均挑战了笛卡尔的几何造型,它们将垂直与水平元素融合为蛇形的树脂表面,并在其上涂了一层极具表现力的金橘色。家具元素在遵循一系列普遍拓扑规则的同时,添加了与众不同的独特设计特色,在与彼此相互呼应、融为一体的同时,保留了各自雕塑般的独立性。每一件家具均展现出多种潜在的功能,创造性地将其表面部分、展示区域和落座元素结合在一起。

卡特桌（CRATER TABLE）
大卫·吉尔画廊，2007年
与帕特里克·舒马赫合作

这是一张将桌子的本体连同其上的物品均被缩减为最纯粹的形式的桌子，最大化地表达了其功能性。虽然凸出于桌面之上的物品看起来各自独立，但当你换个角度观察时，却会发现这些碗以及烛台根本不具备稳固的独立性；它们仅仅是组成元素，存在于桌子那不断延伸的表面上，而在桌面之下，变换出三根凸出物，以构成桌子独具特色的造型。看似无形的力形成了三处凹陷的坑：其中两个延伸至底面，而第三个则悬浮在半空中，后者可以明显看出是一个碗。为了设计出最具动态和神秘特色的表面，我们研究了铝的独特性质，以塑造出流动的造型。

月亮体系沙发（MOON SYSTEM）
B&B意大利（B&B ITALIA），2007年
与帕特里克·舒马赫合作

这套沙发的设计创造了独特的新型坐具的概念。为了设计出这款家具，我们将尖端意大利家具生产商的制作经验融入我们多年以来所做的复杂流线造型研究之中。在这套组合家具中，人体工程学与美感被融入延展的形状之中，通过将每一个元素塑造为单独的模块，我们重新定义了坐具的意义。通过旋转、交织和隐藏个体元素，这款家具系统重新定义了自身的意义，从而在整体的组合中，融合了个体的家具造型，模糊了虚无与真实、积极与消极、客体与空间之间的界限。司空见惯的沙发造型被消解了，从而设计出灵活而舒适的造型，以满足使用者的多种需求。

这张桌子的设计灵感来源于一次建筑实验,当时我们想要设计如雕塑般耸立于城市中的建筑,它们交织在一起形成了群集的整体。我们将桌子这一概念分散为底面、构架和表面,从而在两块水平面板之间营造出了一个世界,二者之间的空隙塑造着实体的造型。这些空隙并非仅仅是孔洞,它们也定义了桌面——如同睡莲一般,平缓的叶面是由其下隐没的复杂结构所支撑的。一反常态,桌面被简化得毫无结构可言,仅仅被分为有机构造的四部分,它们发挥着"餐具垫"的作用。无形的力量拉扯着桌子的一端,使对称的结构变得歪斜,而面积逐渐缩减的船首造型则拉动着其他的构造,桌面下的结构也与之配合着达成扭曲的平衡。因此,这项设计微缩了我们所设计建筑中的某些空间理念。设计中的形状不仅仅发挥着功能层面的作用,同样也展示着这项设计的叙事性及其空间流动感——它们召唤着一个世界,一个由隐形力量和黑色物质所构成的世界。

梅萨桌(MESA TABLE)
维特拉(VITRA),2007年
与帕特里克·舒马赫合作

扎哈·哈迪德枝形吊灯（ZAHA HADID CHANDELIER）

施华洛世奇水晶宫（SWAROVSKI CRYSTAL PALACE），2008年

与帕特里克·舒马赫合作

在自组织系统和纳米技术的启发下，这项为施华洛世奇所设计的作品挑战了枝形吊灯的原有概念，将其重新定义为占据一定空间的物体，而非仅仅是悬挂于天花板上的物品。86根电缆以45度角从地面延伸至天花板，其拉力超过8吨，从而支撑起一根有凹槽的圆锥造型，其内装有2700根照明水晶。水晶体以流动的旋涡造型环绕圆锥造型一周，以浅蓝色的光晕营造出轻妙的氛围。这盏巨大的枝形吊灯长度超过15米，其内部设计富丽堂皇，却也蕴含着高度优雅的内涵。

群组型吊灯（SWARM CHANDELIER）

艾斯塔子弟家具公司，2006年

与帕特里克·舒马赫合作

这张照片拍摄于伦敦设计博物馆的一场展览上。这款吊灯由黑水晶柱组成，它们被塑造成了动态的造型，而非静态的造型。水晶柱划破了空间，在可感知的真实时间中营造出动态的效果。其错综复杂的分层结构并没有遵循任何比例规则，也未曾突出某种对称形式。相反，在收放有度的爆发所产生的动态力量之中，设计元素在空间关系之中融合为整体。吊灯的花瓣状造型被聚集在一起，形成了规模更大的整体。花瓣造型之间并不是互不相关的，它们在不断散发的张力之间，完成了与彼此的呼应。

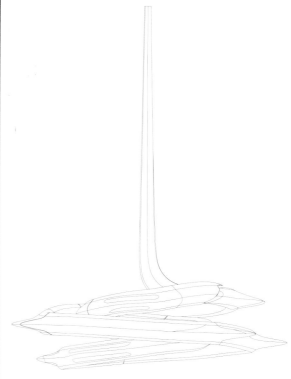

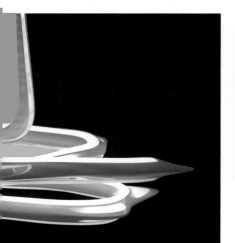

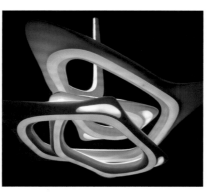

  流动性与流畅性这两个概念术语恰如其分地描绘了这盏宽达1.8米的吊灯。其造型呈双螺旋状,以复杂的弧度上下贯穿,塑造出首尾相接的结构,由此形成了无限延展的条带状吊灯。按照设计规划,这盏吊灯被塑造成了星状造型,造型元素从吊灯的中心向外伸展,由此而突出想象中的离心力。两盏透明的亚克力螺旋灯被配置于原本透明的表面之上。一条凹陷的液晶灯带营造出精心设计的灯光效果。直接照明和间接照明都可以按需关闭,从而营造出不同的照明气氛,以适应吊灯所在的特定空间。作为技术支撑,先进的数码制造方法丰富了崭新的内部设计理念,我们鼓励使用者创造性地探索枝形吊灯的互动性质,感受吊灯与众不同的美学特征。

旋涡吊灯(VORTEXX CHANDELIER)
萨瓦亚和莫洛尼,2005年
与帕特里克·舒马赫合作

冰隙花瓶（CREVASSE VASE）

艾烈希·斯帕（ALESSI SPA），2005—2008年
与帕特里克·舒马赫合作

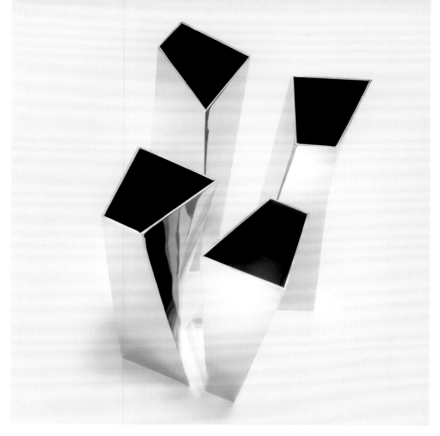

这些花瓶是从一块整体材料上切割下来的，之后我们又沿着两条对角线为它们塑形，从而塑造出了弯曲的表面。我们可以以不同的组合方式将它们组合在一起，组合出一个整体的造型，也可以将它们作为独立的个体而分开摆设。这一套花瓶本身具备令人玩味的性质，人们可以改变其组合形式，由此，可以塑造出各种各样的造型，那感觉就像是在玩一套不断变换的拼图游戏。

完美福餐具（WMF CUTLERY）

完美福（WMF），2007年
与帕特里克·舒马赫合作

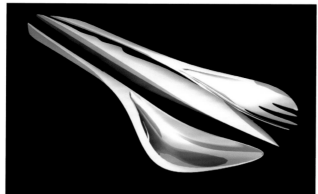

这套造型独特的餐具体现了我们在建筑学的研究与工业设计的探索之间的张力。每一件独立的餐具都郑重地表达着自身的内涵，然而，当它们被组合在一起的时候，作为一个统一体，它们彼此之间呼应着。雕塑般的表面不断延展，构成了复杂的集合物体，通常用于汽车设计的递升曲线被应用到这套餐具的设计之中，从而赋予了它高度的动态效果。这套餐具的材料为坚固的不锈钢，我们将其抛光为镜面一样的效果。在餐具的某些部位，其厚度达到极致，从而使这套餐具中的每一件都与彼此达成最大限度的平衡。

ZH系列门把手 (SERIES ZH DOOR HANDLES)

瓦利和瓦利 (VALLI & VALLI),2007年

与姚伍迪合作

我们为马德里的美国普尔塔酒店(见第155页)精心设计了这款门把手,以用于其公共空间和一楼的房间。高度自定义的坚硬结构丰富了这款设计的特色。我们竭尽全力将门把手材料的冰冷融入了设计触感的流畅性,并在充分考虑空间结构的前提下,确定了门把手的首端和末端部位。光滑的流线有机融合为和谐的整体,塑造出美感与复杂相结合的立体效果。

艾康包(ICONE BAG)
路易威登(LOUIS VUITTON), 2006年
与帕特里克·舒马赫合作

为路易威登所设计的这款标志性"水桶"包为我们提供了以此重新阐释的机会,它引导着我们思考手提包作为普通容纳性器物的功能。在这款设计中,我们将形状层面的诸多操作(挤压、扭曲、剥离和分层)与选材结合在一起,设计出一系列形态各异的手提包,从而创造出与传统手提包造型相结合的类型(袖珍手提袋、无带提包和水桶包)。我们将这些类型相融合,由此创造出诸多潜在的关系,从而让使用者能够在一致的组合形式中替换箱包的部件。使用者可以组合出不同的造型,由此激发了使用者与箱包之间的互动,在不同的情况下,水桶包的形状与功能均能按需组合。

梅利莎鞋(MELISSA SHOE)
梅利莎和格林迪恩公司(MELISSA, GRENDENE S/A), 2008年
与帕特里克·舒马赫合作

与时尚界的合作为我们带来了一个激动人心的机会,使我们能够以不同的规格,通过不同的途径表达空间概念。这款设计专注于身体的流动轮廓,而鞋子不对称的造型传达了内在的动态韵味,从而唤起了接连不断的动态效果。设计理念强调鞋子的动态韵味,而非将其构造成一件用于商店橱窗展示的静态作品。这双鞋子立于地面之上,以温柔优雅的姿势爬上脚踝,有机塑料使这双鞋温柔地贴合在皮肤上。一种内敛的轻盈模糊了身体与物体之间的界限。

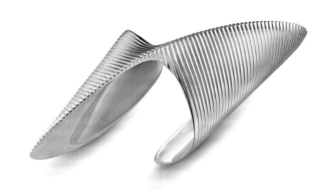
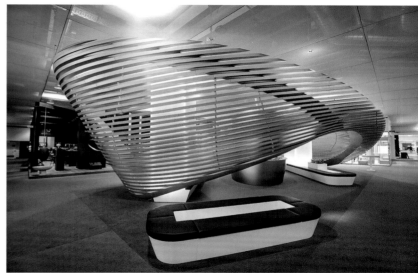

片体组合（LAMELLAE COLLECTION）

乔治·杰生（GEORG JENSEN），2016年
与帕特里克·舒马赫合作

这套为乔治·杰生所设计的片体9件套装包含5件戒指和3件手镯，它们于2016年被首次展览于巴塞尔世界珠宝展览上。我们使用三维的设计与制作工艺，设计并雕饰了这些首饰。一部分首饰由纯银打造，另一部分首饰由黑金线打造，其上镶嵌着黑钻石。

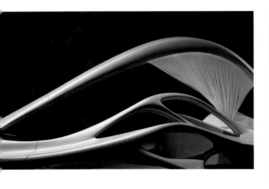

莲花,威尼斯双年展(LOTUS, VENICE BIENNALE)

意大利,威尼斯,2008年
与帕特里克·舒马赫合作

这套装置是由若干个部分组合而成的,在不同的场合中,可以根据使用者的需要组合或者是分拆,以供人们落座休息,或者是发挥其存储功能。诸多折叠结构不仅仅赋予了它们以独特的造型,而且使其具备各种各样的功能。空间在两种极端状态之间变动:一种为压缩状态,从周围的环境中脱离出来;另一种为展开、分散的状态,与周围的环境交织为一体。随着这套家具被拆分为桌椅、床、衣架、衣柜和茶几,可移动的家具与建筑便可融为一体。在这项设计中,一件件嵌入式家具或被有意遮挡,或欲盖弥彰,从而为处所提供了多种多样的可能性。

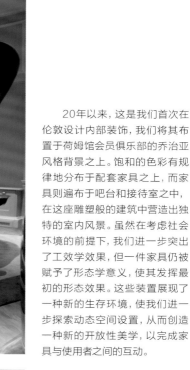

荷姆吧台（HOME BAR）
英国，伦敦，2008年
与帕特里克·舒马赫合作

20年以来，这是我们首次在伦敦设计内部装饰，我们将其布置于荷姆馆会员俱乐部的乔治亚风格背景之上。饱和的色彩有规律地分布于配套家具之上，而家具则遍布于吧台和接待室之中，在这座雕塑般的建筑中营造出独特的室内风景。虽然在考虑社会环境的前提下，我们进一步突出了工效学效果，但一件家具仍被赋予了形态学意义，使其发挥最初的形态效果。这些装置展现了一种新的生存环境，使我们进一步探索动态空间设置，从而创造一种新的开放性美学，以完成家具与使用者之间的互动。

光晕（AURA）
意大利，威尼斯，2008年
与帕特里克·舒马赫合作

为了纪念帕拉迪奥的500年诞辰，我们选择以佛斯卡利别墅（也被称为玛尔孔滕塔酒店）的一间单间为基础，探索其逻辑与关联系统。这座别墅由帕拉迪奥建于1555年，意在论证他的建筑理论。我们所引入的动态构成元素撼动了这座建筑的自然平衡，摒弃了由欧几里得数学原理所衍生的帕拉迪奥比例理论，以探索先进数码技术的潜力。由此，这项设计被赋予了空间形态学意义，以展现缥缈空间中的空隙结构以及骨架结构。

藤蔓（CIRRUS）

美国，俄亥俄州，辛辛那提，2008年
与帕特里克·舒马赫合作

这款"地毯式"内部装修是为洛伊斯和理查德·罗森塔尔当代艺术中心（第97页至第99页）所设计的，它呈条纹编织状沿墙壁一直蔓延至地板。稠密的空隙环绕出这一特殊造型，为使用者提供了落座和倚靠的区域。这些生动的雕塑造型遵循着流动而多孔的建筑语言，在这片编制的造型中，每一束独特的条纹都是极其重要的元素。

首尔办公桌和餐桌（SEOUL DESK AND TABLE）

纽约项目（NEW YORK PROJECTS），2008年
与帕特里克·舒马赫合作

这项设计将结构元素融合为流线型的造型，从而探索了水平与垂直之间转换的瞬间。我们利用先进的汽车设计以及制造技术，以复杂的工艺将碳纤维塑造为桌椅，从而赋予了这些家具以超乎寻常的轻度和韧性，并加之以独具特色的纹理装饰。

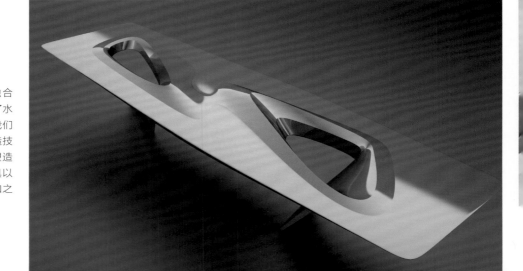

斯图亚特·韦茨曼精品店（STUART WEITZMAN BOUTIQUES）

米兰，香港，罗马，2013—2014年
与帕特里克·舒马赫合作

这项设计理念突出了几何结构之间的嬉戏与互动，意在以褶皱和凹陷营造出韵律感。我们设计了中心展示装置——由镀玫瑰色的玻璃纤维制成，以陈列斯图亚特·韦茨曼的收藏品，并为参观者们提供了座位。我们将这些元素并置，展示这间精品店的与众不同。我们在墙体和天花板上使用了玻璃纤维加固混凝土，这不仅仅具备坚固的特性，而且还可以精确地塑造出复杂的弧度。置身于诸多微妙的单色阴影所调制出的色彩之中，这项设计营造出一片待顾客去探索的风景，从而深化了顾客与陈列品之间的互动。

桌子、架子和亨利·摩尔展览设计（TABLE, SHELF AND HENRY MOORE EXHIBITION DESIGN）

霍瑟与沃斯（HAUSER & WIRTH），2008年
与帕特里克·舒马赫合作

为了展览亨利·摩尔那罕见的小型雕塑，我们设计了这款作品。我们应用了最新的汽车制造技术，将坚实的铝块削减为基本的造型，随后技艺精湛的匠人将这些造型焊接为一体，手工抛光，以制作出浑然一体的物品。纤维像一面帆布一般延展于支架之上，形成了一面面墙壁。

金勺沙发（SCOOP SOFA）

萨瓦亚和莫洛尼，2008年
与帕特里克·舒马赫合作

这款沙发被设计为流动的线性造型，它延续了40多年以来我们所追求的造型设计风格。虽然一眼望去这件作品毫无雕饰、浑然一体，但设计体却遵循着我们所研究的至高形式逻辑，仿佛游离于持续转换和平稳过渡的体系之中。这项设计延续了以往的风格，诸如Z-斯凯珀（第264页）、皮亚扎茶与咖啡用具（第263页）、浅绿色桌子（第269页）以及美国普尔塔酒店（第155页）。在这项设计中，我们建立了一种游离于桌面反光之外的空间关系。

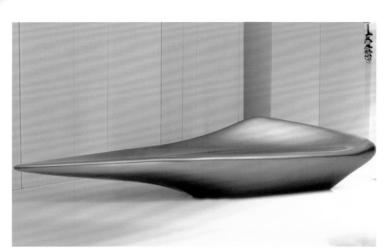

如同一朵绽放的抽象花儿一般,这座户外雕塑从景观中脱颖而出。它的动态似乎凝结在了时间之中,为我们提供了夸张的悬吊式座椅。我们采用了最先进的电脑控制磨边技术来完成这件作品,最终的成品超越了最初的设计。其外壳由纤维玻璃制成,其上的镀层营造出一种新颖的镜面效果,同时嵌入和反射着周围的环境。

克罗莉丝(KLORIS)
罗孚画廊(ROVER GALLERY),2008年
与帕特里克·舒马赫合作

精确切割工艺制作出的水晶纯度和水晶周围所巧妙环绕的有机形状,这二者之间的张力在这套新首饰套系中得到了充分的体现。这套首饰的灵感来源于自然中的蛇状造型,体现了我们在设计以及将设计变为现实的每一阶段中所进行的持续实验和创造。每一颗小水晶都浓缩了一瞬的时间,将时间定格于凝滞状态,将张力展现于生物形态造型的流动性与水晶那精致的规律性之间。每一颗水晶的表面弧度都是迂回而曲折的,其动态的形状与人体工程学天衣无缝地融为一体。不对称的优雅设计中透露着内敛的动态,将内嵌的水晶构件极富魅力地包裹于其中,而轻质的有机材料则转换着水晶所折射的光线。若干构件被排列成动态的造型,进一步提升了该造型设计的优雅度。

冰晶组合(GLACE COLLECTION)
施华洛世奇(SWAROVSKI),2009年
与帕特里克·舒马赫合作

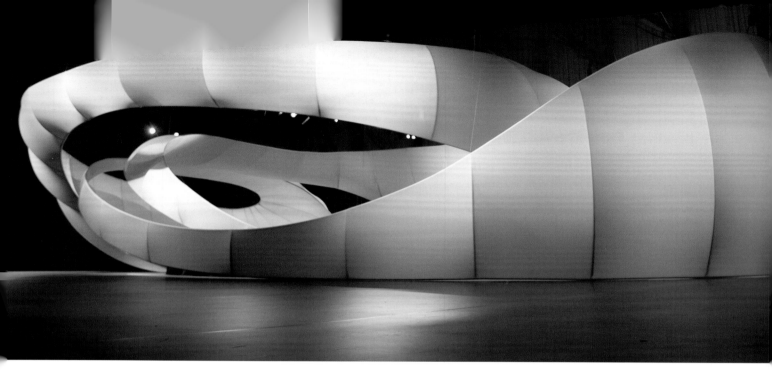

约翰·塞巴斯蒂安·巴赫音乐厅 (JOHANN SEBASTIAN BACH MUSIC HALL)

英国，曼彻斯特，2009年
与帕特里克·舒马赫合作

这项独特的配置位于曼彻斯特艺术馆之内，是专门为约翰·塞巴斯蒂安·巴赫的音乐所设计的。这条卷曲的绸带状造型由透明的纤维薄膜制成，其内部的钢筋结构将其悬挂在天花板上，使其盘旋于室内，在视觉层面上与作曲家形成空间回应，从而达成二者之间错综复杂的和谐关系。绸带造型倾斜在作曲家上方的空间中，如瀑布一般倾斜至地面，随后将听众包裹于其中，而房间也随之被切割为诸多流动的空间，它们彼此之间也处于相互盘绕、吞没、交织的动态之中。缎带造型绕着自身一圈圈扩展开来，从而构造了一层又一层的功能，先是将其自身压缩至旋转楼梯栏杆的规模，随后又舒展为从地板至天花板的高度。纤维的表面在内部结构中蔓延，持续变换并有节奏的起伏波动着，时而构造着绸带造型外部那紧绷的外皮，时而营造出绸带造型内部那柔软而汹涌的效果。透明的亚克力吸音板悬浮于舞台之上，它们一边阻挡并分散着声音，一边隐没于薄薄的织物之中。

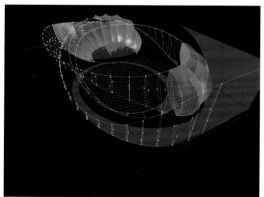

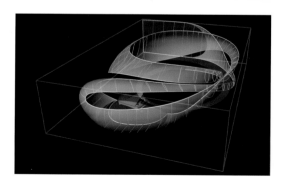

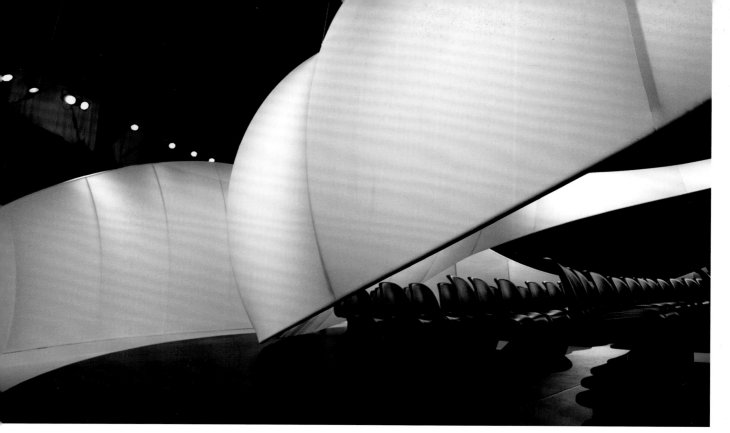

阿维林催弗洛水龙头（AVILION TRIFLOW TAPS）

阿维林催弗洛（AVILION TRIFLOW），2009年

与帕特里克·舒马赫合作

水的流动形式为我们提供了灵感，让我们设计出这一款具有革命性的用于厨房和洗手间的新式水龙头。我们运用了最新的快速成型技术，配合陶瓷取心工艺，每一个水龙头都是定制的艺术品。用于过滤自来水的管道由触敏电子驱动器控制，一根控制杆同时控制冷水和热水，这些均被纳入了水龙头的造型之内。

绞丝手镯（SKEIN SLEEVE BRACELET）

萨伊格珠宝商（SAYEGH JEWELLER），2009年

与帕特里克·舒马赫合作

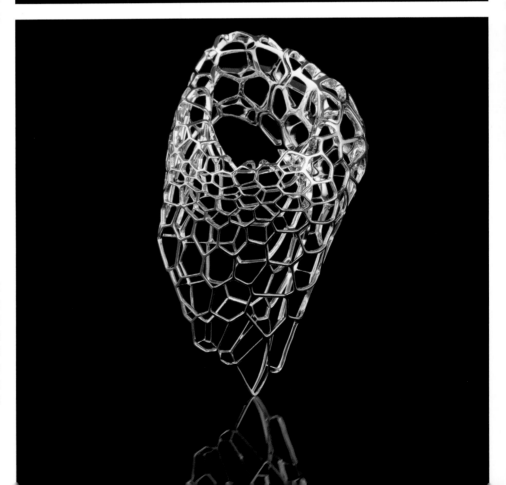

一位叙利亚珠宝商委托我们设计一只多面手镯，我们应用金银丝细工艺将金银丝拉伸延展为格子造型，使其能够环绕于手腕之上。抛光的白金提升了手镯造型精制的复杂性，构成了手镯的格子状造型，而白色的钻石则点缀于手镯之上，如同一颗颗流动的水珠。

我们为阿尔泰米德设计了这盏灯,其灵感来源于我们对自然界生长系统的探索。如同树丛之上的树叶天篷,在这项设计中,横扫的天篷结构从底座处向上延伸,而底座则由相互交错的网状支撑结构构成。二者之间的有机相似性存在于基本的构成元素之中,我们将其由抽象的范式转换为流线的造型。随着这盏灯的结构从底部上升,其结构的复杂性也相应提升。如同一株生长的有机体一般,中心支撑处延伸展出枝状结构,以放射性的姿态延展为规模更大的形状,从而提升了该结构的动态张力。在自然的进一步启发下,我们用这些枝状结构圈框下空隙,以提高主动与被动、凸起与凹陷、扩张与挤压之间的对比。枝状结构遵循着形态学的造型逻辑,从中心处向外延伸以支撑光源。

吉尼丝灯(GENESY LAMP)
阿尔泰米德(ARTEMIDE),2009年
与帕特里克·舒马赫合作

我们在设计这款架子的时候,参考了极小曲面的几何学和数学原理,将其表面的平均曲率设置为零(通过将线框浸入皂溶液之中构成薄层,可以制成面积最小的曲面实体模型)。莫比乌斯带和克莱因瓶(Klein bottle)是广为人知的非定向最小曲面设计案例。在这项设计中,我们运用单个模块的几何特性对作品进行大的变化与组合,利用模块自身的结构以及几何特性对其进行装饰。

潮汐(TIDE)
马吉斯(MAGIS),2010年
与帕特里克·舒马赫合作

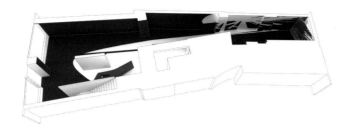

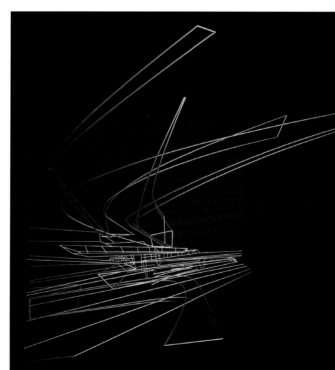

扎哈·哈迪德：建筑师与至上主义（ZAHA HADID: ARCHITECTS & SUPREMATISM）

瑞士，苏黎世，2010年
与帕特里克·舒马赫合作

这场创意展览与第41届巴塞尔艺术博览会同时举行，于是我们的作品与俄罗斯先锋派的作品被并置展览。我们将二维的画作投射至三维的空间之中，从而设计出这一幅动态的黑白设计作品。画廊本身在效果上就是一幅空间画作，画板已被延伸，我们可以在画作中来去穿梭。

我们从俄罗斯先锋派的作品中获得了灵感，尤其是卡济米尔·马列维奇（Kazimir Malevich）的作品。他站在这里捕捉着瞬间的抽象发现，并将其作为启发式原则，以推动创造性工作至前所未有的创新层面。他摒弃了模仿的方式，任凭不受束缚的创造力在无限容纳的空白画布上自由地施展。空间，甚至是世界本身，都成了纯粹创作的场所。我们的作品表现了马列维奇、埃尔·利西斯基（El Lissitzky）和亚历山大·罗钦可（Alexander Rodchenko）绘画以及雕塑作品中的扭曲空间和反重力空间，将其转化为建筑语言。

扎哈·哈迪德：帕多瓦法理宫
(ZAHA HADID: PALAZZO DELLA RAGIONE PADOVA)
意大利，帕多瓦，2009—2010年
与帕特里克·舒马赫合作

我们曾探索并实验了数码设计和构造原则，这次回顾展便检验了我们所做的探索和实验。在这次展览设计中，我们致敬了中世纪帕多瓦法理宫的内部空间以及环境特色，与此同时我们引入了自身所掌握的数码和液态流动技术，从而将空间整合为一体的流动风景，其中遍布着单个的碎片和组合的群体。高低不同的展示模块构成了一个体系，遍布于展厅之中，在遵循展厅整体性的同时，赋予了崭新的意境。展品被摆放于每一个模块之上，按照设计主题，各种各样的作品被排列于展厅之中，其中包括素描、绘画、研究模型、设计原型和视频。

乐家伦敦展厅 (ROCA LONDON GALLERY)
英国，伦敦，2009—2011年
与帕特里克·舒马赫合作

这间全球乐家展厅中的最新标志性展厅位于伦敦西部的皇家码头。在这项设计中，水流发挥着主要的作用。水流畅通无阻地流淌于物体表面，以侵蚀作用构造出内部空间，随后流过主展厅。在水流的侵蚀作用下，建筑表面出现了若干裂口，而建筑的内部则布满了反光的水珠，从而为展厅内的艺术作品提供了背景。这些水珠将空间的不同区域连为一体，其中包括一间会议室、一间咖啡厅和酒吧、一间图书馆、一面多媒体墙、一间接待室和许多视频屏幕。展厅内部使用了最新技术，为了向参观展厅的顾客提供真正前沿的体验，我们还配备了视听、音响和灯光设备。

Z-椅子（Z-CHAIR）

萨瓦亚和莫洛尼，2011年
与帕特里克·舒马赫合作

在这款设计中，简单的三维造型在空间中呈蛇形蜿蜒着，表现着形式与功能、优雅与实用、差异性与连续性之间永无止境的对话。其流线型的环状构造中透出抽象的几何意义，在蔓延的轨迹中诉说着自身的语言。单薄的流线与宽阔的立面相互交替，从而为整个设计造型提供了内在的稳定性，散发着人体工程学的智慧意蕴。尖锐的拐角与顺畅的流线之间形成了微妙的对比，从而调和了优雅的造型与张扬的表现力。

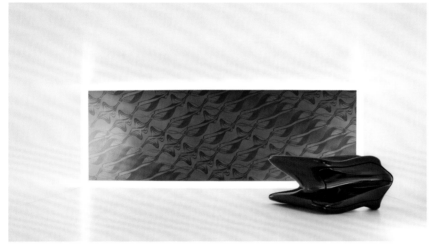

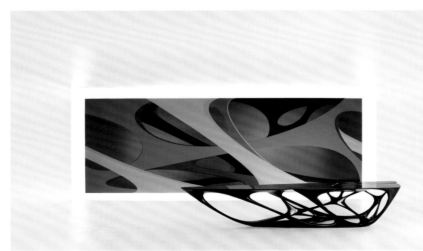

扎哈·哈迪德的艺术边缘（ART BORDERS BY ZAHA HADID）

玛堡壁纸（MARBURG WALLCOVERINGS），2010年

与帕特里克·舒马赫合作

我们为玛堡壁纸公司设计了四款作品（"蜂窝式""条纹式""漩涡式"和"弹簧式"），构成了其"艺术边缘"系列壁纸的一部分。这次众望所归的合作十分振奋人心，在合作过程中，我们将技术创新融入了艺术视野之中。该项目启动于法兰克福展览会——国际家纺展览会期间，这些设计以生动的空间性与流动性打破了常规。一款独具特色的设计宽达9米、高达3.3米，通过变换和压缩空间，为房间营造出了动态的效果。壁纸之间相互填充，首尾相接，将墙体抽象为无尽的画布，以深邃的意境展现着动态的视觉效果。

扎哈·哈迪德：流动性与设计
(ZAHA HADID: FLUIDITY & DESIGN)

巴林，穆哈拉格，2010年
与帕特里克·舒马赫合作

在这次展览中，我们进一步探索了流动的新建筑语言，将其拓展至所有规模的设计，并以此展现了空间的复杂性。同样，我们也借此机会以激进的方式重新阐释了空间。我们的展览规划遵循了本·玛塔大厦（Bin Matar House）内部的正交性质以及组织原则，与此同时，我们也将自身所秉持的有机流动性融入设计之中。展览上所展出的作品包罗万象，小到专门定做的珠宝，大到大型的家具等。

扎哈·哈迪德：一座建筑
(ZAHA HADID: UNE ARCHITECTURE)

法国，巴黎，2011年
与帕特里克·舒马赫合作

2010年，香奈儿将其移动艺术馆（第212页至第215页）捐献给了阿拉伯世界研究院。此举不仅使这座艺术馆永久地留在了巴黎的中心，而且还腾出了富余的空间举办展览。在这一新展区所举办的首场展览是一场主题性探索，探索了我们近年来的研究主题。访客们应邀前来，从三个层面上体验这场展览：探索移动艺术馆（建筑）、观看展场设计（配景设计）、欣赏我们的作品（展品）。

在费城艺术博物馆的这场展览中，我们营造了包罗万象的环境，以展览近年来我们所设计的家具和物件样品，其中包括梅萨桌（第275页）以及第一代Z-汽车的原型（第270页）。这场展览首先在美国举行，我们在哈迪德亲自设计的背景中突出了她的作品。展厅的造型中包含起伏的支架，是由成品聚苯乙烯塑造的，其上遍布着基于流线型几何结构的乙烯基图形。在这项展览设计中，我们探索了基于流线型动态的造型语言，强调了我们所设计作品中惯有的特质，以及建筑、都市生活和设计是如何紧密结合在一起的。

扎哈·哈迪德：运动的形态（ZAHA HADID: FORM IN MOTION）

美国，宾夕法尼亚州，费城，2011—2012年

与帕特里克·舒马赫合作

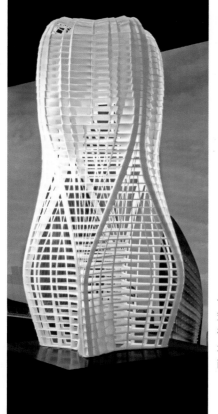

在德国科隆AIT建筑沙龙的展览上，我们所展出的一系列设计突出了我们所做的参数化研究。用来表现这一新兴类型的媒介，包括模型、绘画和多媒体演示手段。在参数化研究中，我们探索了自己的作品和实验的多样性。从小规模的设计物件，到大型的城市规划概念，均折射出我们想要囊括所有规模的巨大野心。我们为这座建筑设计了专门的布景，地板、墙壁和天花板均隐没于配景之中，而内部空间也随之发生了变化。白色的混凝土表面让位于黑色的纤维、白色的薄膜、银色的墙纸和跃动的投影。

扎哈·哈迪德：参量塔研究
(ZAHA HADID: PARAMETRIC TOWER RESEARCH)

德国，科隆，2012年

与帕特里克·舒马赫合作

意大利陶瓷制造商（Lea Ceramiche）委托我们设计了这件作品，我们借此机会在当代重新阐释了米兰大学的18世纪院落，将生硬的笛卡尔图形转换为线性流动的动态空间。包括扁平的瓷砖在内，我们在这项设计中运用了复杂的三维弧线几何形状，以提升不同元素的雕塑感，并赋予它们动态的造型，从而达成实体与空隙之间的平衡关系。每一件单独的物件不仅仅可以被看作是整体的一部分，而且还可以被当作游离于引力场中的碎片。

旋转（TWIRL）
意大利，米兰，2011年
与帕特里克·舒马赫合作

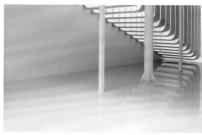

我们将这款楼梯设计为悬浮的样式，意在以此展现轻盈的展览空间。每一节悬浮的楼梯都被设计为一条独立的带状造型，它们一节一节地向下排列着形成了管状楼梯。这些带状造型由性能极高的混凝土构成，该材料具备很高的强度，从而使带状造型能够保持相对轻盈的结构。意大利的著名船厂（Il Cantiere）为我们制作了单体可调节模具，每一级台阶均是从这一模具中制造出来的。随着台阶不断地向下延伸，每一节台阶都变得越来越狭窄。我们所设计的这组楼梯是可以拆卸的。

浮动的楼梯（FLOATING STAIRCASE）
英国，伦敦，2012年
与帕特里克·舒马赫合作

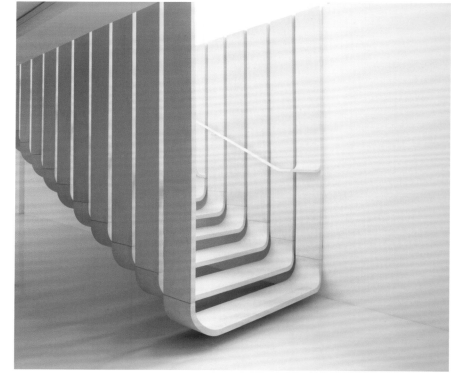

阿鲁姆装置与展览 (ARUM INSTALLATION AND EXHIBITION)

意大利，威尼斯，2012年
与帕特里克·舒马赫合作

我们为威尼斯双年展所设计的这款装置由褶皱的金属构成，我们希望以此致敬费雷克斯·堪迪拉（Félix Candela）、海恩兹·伊斯勒（Heinz Isler）和弗雷·奥托（Frei Otto）的开创性作品，他们在探索形式的过程中完成了优雅的设计。同样，我们也要感谢包括菲利普·布洛克（Philippe Block）在内的现代学者们，他们的作品中包括石材压缩的壳状造型。我们希望探索轻质壳状造型这一独特的领域，与此同时赋予该造型延展性的结构。我们已经设计了许多复杂的壳状造型，也已经设计了许多延展性结构。而这一次，我们终于尝试将二者结合起来。

CITCO
ZAHA HADID
COLLECTION

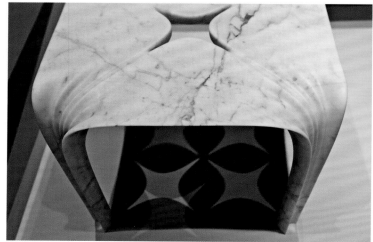

西特科大理石花瓶和四方桌（CITCO TAU VASES AND QUAD TABLES）
意大利，米兰，2015年
与帕特里克·舒马赫合作

　　这一系列的花瓶和桌子是为意大利西特科大理石公司设计的。在这项设计中，哈迪德以她那独一无二的方式重新诠释了大理石。大理石花瓶的构造十分自然，褶皱复杂地聚拢在一起，虽然它们的构成材料非常坚固，却给人一种易碎的错觉。我们在四方桌的中央部位设计了孔隙，桌面从孔隙处向四周延展，而微妙的褶皱则在桌腿顶端聚拢，它们向下延伸直至桌腿的末端。我们为每件物品都设计了五种不同的造型。

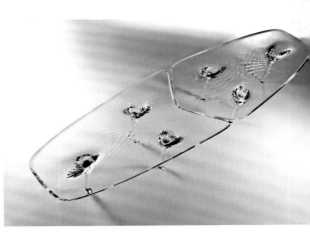

在这项设计中，透明桌面之下是曼妙的波浪和涟漪，似乎要将平滑的桌面上静态的几何结构转换为动态的效果，而桌腿则似乎以旋涡的形式从桌面之上倾泻而下，时间便永恒地凝固在这一刻。透明的新型材料强化了这一效果，如千变万化的万花筒一般，折射着无暇而深邃的光，提升了整个设计的深度和复杂性。最终，我们营造出动态的表面效果，随着使用者视角的变化而不断地发生改变。这款设计在具备更重功能、满足人体工程学需求的同时，使我们探索了空间中的动态并取得了明显的进展。

流体冰餐桌（LIQUID GLACIAL DINING & COFFEE TABLE）
大卫·吉尔画廊，2012年
与帕特里克·舒马赫合作

扎哈·哈迪德回顾展
(ZAHA HADID RETROSPECTIVE)

俄罗斯,圣彼得堡,国立艾尔米塔什博物馆
2015年6月27日—9月27日

　　这是哈迪德在俄罗斯所举办的第一届作品回顾展,该展览强调了俄罗斯先锋派对其设计的影响,尤其是在其早期设计生涯之中。这次展览举办于冬宫的尼古拉耶夫斯基历史展厅之中,我们从哈迪德40年以来的全部作品中挑选出开创性的绘画、素描、模型和设计物品,将其以各种各样的媒介形式展览在观者眼前,其中包括电影、照片以及配套设备等。

　　"先锋派极富冒险精神,他们敢于拥抱新事物,极其崇拜创造能力,他们的这些特质吸引了我,"2004年,当哈迪德在国立艾尔米塔什博物馆接受普利兹克建筑奖时曾这样说,"我们获得了从前所难以想象的创造自由,抽象艺术取得了巨大的突破,由此带动了现代建筑业的发展,我领悟了现代艺术发展的诀窍。我们要具备这样的理念,空间本身能够被扭曲和变形,在保持连贯性和持续性的前提下,空间也可以具备动态性与复杂性。"

项目信息

Malevich's Tektonik [18]
London, UK, 1976–1977
Fourth-year Student Project

Dutch Parliament Extension [19]
The Hague, Netherlands, 1978–1979
Design Team Office for Metropolitan Architecture (OMA): Zaha Hadid, Rem Koolhaas, Elia Zenghelis, with Richard Perlmutter, Ron Steiner, E. Veneris

Museum of the Nineteenth Century [19]
London, UK, 1977–1978
Fifth-year Student Design Thesis

Irish Prime Minister's Residence [20]
Dublin, Ireland, 1979–1980
Design Team Zaha Hadid with K. Ahari, Jonathan Dunn

Parc de la Villette [21]
Paris, France, 1982–1983
Design Team Zaha Hadid with Jonathan Dunn, Marianne van der Waals, Michael Wolfson

59 Eaton Place [21]
London, UK, 1981–1982
Design Team Zaha Hadid with Jonathan Dunn, K. Knapkiewicz, Bijan Ganjei, Wendy Galway

The Peak [22–23]
Hong Kong, China, 1982–1983
Design Team Zaha Hadid with Michael Wolfson, Jonathan Dunn, Marianne van der Waals, N. Ayoubi
Presentation Michael Wolfson, Alistair Standing, Nan Lee, Wendy Galway
Structural Engineer Ove Arup & Partners: David Thomlinson

The World (89 Degrees) [24]
1983
Painting

Melbury Court [24]
London, UK, 1985
Design Team Zaha Hadid with Brian Ma Siy, Michael Wolfson

Grand Buildings, Trafalgar Square [25–27]
London, UK, 1985
Design Team Zaha Hadid with (in the early stages) Brian Ma Siy
Competition Team Michael Wolfson, Brian Ma Siy, Marianne Palme, Kar Hwa Ho, Piers Smerin

Halkin Place [28]
London, UK, 1985
Design Team Zaha Hadid with Brian Ma Siy, Piers Smerin

Kyoto Installations [28]
Kyoto, Japan, 1985
Installations

Tents and Curtains, Milan Triennale [29]
Milan, Italy, 1985
Design Team Zaha Hadid with Piers Smerin, Michael Wolfson

24 Cathcart Road [29]
London, UK, 1985–1986
Client Bitar
Design Team Zaha Hadid with Michael Wolfson, Brett Steele, Nan Lee, Brenda MacKneson

Hamburg Docklands [30]
Hamburg, Germany, 1986
Masterplanning Workshops

New York, Manhattan: A New Calligraphy of Plan [31]
1986
Painting

Kurfürstendamm 70 [32]
Berlin, Germany, 1986
Client EUWO Holding AG
Design Team Zaha Hadid with Michael Wolfson, Brett Steele, Piers Smerin, Charles Crawford, Nicola Cousins, David Gomersall
Client Feasibility Berlin Senate
Co-architect Stefan Schroth
Structural Engineer Ove Arup & Partners: Peter Rice, John Thornton
Glazing Consultant RFR Engineers: Hugh Dutton
Quantity Surveyor Büro am Lützowplatz: Wilfraed Kralt
Total Area 820 m² (7 floors)

IBA-Block 2 [33]
Berlin, Germany, 1986–1993
Client Degewo AG
Design Team Zaha Hadid with Michael Wolfson, David Gomersall, Piers Smerin, David Winslow, Päivi Jääskeläinen
Co-architect Stefan Schroth
Total Area 2,500 m² (long block: 3 floors; tower: 8 floors)

Azabu-Jyuban [34]
Tokyo, Japan, 1986
Client K-One Corporation
Design Team Zaha Hadid with Michael Wolfson, Brenda MacKneson, Alistair Standing, Signy Svalastoga, Paul Brislin, Nicola Cousins, David Gomersall, Edgar González, Erik Hemingway, Simon Koumjian, Päivi Jääskeläinen
Models Daniel Chadwick, Tim Price
Project Architect (Japan) Satoshi Ohashi
Co-architect Hisashi Kobayashi & Associates
Structural Engineer Ove Arup & Partners: Peter Rice, Yasuo Tamura
Total Area 340 m² (6 floors)

Tomigaya [35]
Tokyo, Japan, 1986
Client K-One Corporation
Design Team Zaha Hadid with Michael Wolfson, Brenda MacKneson, Alistair Standing, Signy Svalastoga, Paul Brislin, Nicola Cousins, David Gomersall, Edgar González, Erik Hemingway, Simon Koumjian, Päivi Jääskeläinen, Patrik Schumacher
Models Daniel Chadwick, Tim Price
Project Architect (Japan) Satoshi Ohashi
Co-architect Hisashi Kobayashi & Associates
Structural Engineer Ove Arup & Partners: Peter Rice, Yasuo Tamura
Total Area 238 m² (2 floors)

West Hollywood Civic Center [36]
Los Angeles, California, USA, 1987
Administrative Buildings

Al Wahda Sports Centre [37]
Abu Dhabi, UAE, 1988
Client Sheikh Tahnoon bin Saeed Al Nahyan
Design Team Zaha Hadid with Michael Wolfson, Satoshi Ohashi
Structural Engineer Ove Arup & Partners: Peter Rice

Metropolis [38]
1988
Painting

Berlin 2000 [39]
1988
Painting

Victoria City Areal [40–41]
Berlin, Germany, 1988
Client City of Berlin (Building Authority)
Design Team Zaha Hadid with Michael Wolfson, Nicholas Boyarsky, Patrik Schumacher, Edgar González, Paul Brislin, Nicola Cousins, Signy Svalastoga, C.J. Lim, Kim Lee Chai, Israel Numes, Mathew Wells, Simon Koumjian
Model Daniel Chadwick
Structural Engineer Ove Arup & Partners: Peter Rice, Mathew Wells
Total Area c. 75,000 m² (15 floors)

A New Barcelona [42]
Barcelona, Spain, 1989
Design Team Zaha Hadid with Patrik Schumacher, Simon Koumjian, Edgar González

Tokyo Forum [43]
Tokyo, Japan, 1989
Client Tokyo Metropolitan Government
Design Team Zaha Hadid with Brian Ma Siy, Patrik Schumacher, Vincent Marol, Philippa Makin, Bryan Langlands, David Gomersall, Jonathan Nsubuga
Model Daniel Chadwick
Total Area 135,000 m² (8 floors)

Hafenstraße Development [44–45]
Hamburg, Germany, 1989
Client Free and Hanseatic City of Hamburg (Building Authority)
Design Team Zaha Hadid with Patrik Schumacher, Signy Svalastoga, Edgar González, Bryan Langlands, Philippa Makin,

Nicola Cousins, Mario Gooden, Ursula Gonsior, Claudia Busch, Vincent Marol
Model Daniel Chadwick
Co-architect Mirjane Markovic
Structural Engineer Ove Arup & Partners: Peter Rice
Total Area Corner building: 871 m² (8 floors); middle site building: c. 2,800 m² (10 floors)

Monsoon Restaurant [46–47]
Sapporo, Japan, 1989–1990
Client JASMAC Corporation
Design Team Zaha Hadid with Bill Goodwin, Shin Egashira, Kar Hwa Ho, Edgar González, Bryan Langlands, Ed Gaskin, Yuko Moriyama, Urit Luden, Craig Kiner, Dianne Hunter-Gorman, Patrik Schumacher
Consultants Michael Wolfson, Satoshi Ohashi, David Gomersall
Model Daniel Chadwick
Producer Axe Co., Ltd
Total Area 435 m² (2 floors)

Osaka Folly, Expo 1990 [48]
Osaka, Japan, 1989–1990
Client and Sponsor Fukuoka Jisho Co., Ltd
Organizer Workshop for Architecture & Urbanism
Design Team Zaha Hadid with Edgar González, Urit Luden, Satoshi Ohashi, Kar Hwa Ho, Patrik Schumacher, Voon Yee-Wong, Simon Koumjian, Dianne Hunter-Gorman, Nicola Cousins, David Gomersall
Model Daniel Chadwick
General Producer Arata Isozaki
Contractor Zenitaka Corporation
Total Area 435 m²

Leicester Square [49]
London, UK, 1990
Client *Blueprint* magazine
Design Team Zaha Hadid with Graham Modlen, Vincent Marol, Simon Koumjian, Patrik Schumacher, Craig Kiner, Cristina Verissimo, David Gomersall, Philippa Makin, Dianne Hunter-Gorman, Maria Rossi, Mya Manakides

Zollhof 3 Media Park [50–51]
Düsseldorf, Germany, 1989–1993
Client Kunst-und Medienzentrum Rheinhafen GmbH
Design Zaha Hadid
Project Architects Brett Steele, Brian Ma Siy
Project Team Paul Brislin, Cathleen Chua, John Comparelli, Elden Croy, Craig Kiner, Graeme Little, Yousif Albustani, Daniel R. Oakley, Patrik Schumacher, Alistair Standing, Tuta Barbosa, David Gomersall, C.J. Lim
Models Ademir Volic, Daniel Chadwick
Feasibility and Competition Michael Wolfson, Anthony Owen, Signy Svalastoga, Edgar González, Craig Kiner, Patrik Schumacher, Ursula Gonsior, Bryan Langlands, Ed Gaskin, Yuko Moriyama, Graeme Little, Cristina Verissimo, Maria Rossi, Yousif Albustani
Consultant Architect Roland Mayer
Project Manager Vebau GmbH
Project Co-ordinator Weidleplan Consulting GmbH
Structural Engineers Boll und Partner; Ove Arup & Partners
Services Engineers Jaeger, Mornhinweg und Partner; Ove Arup & Partners; Ingenieurbüro Werner Schwarz GmbH
Façade Consultant Institut für Fassadentechnik
Fire Specialist Wilfred Teschke

Building Physicist Dr Schäcke & Bayer GmbH
Traffic Consultant Waning Consult GmbH
Cost Consultant Tillyard GmbH

Vitra Fire Station [52–55]
Weil am Rhein, Germany, 1990–1994
Client Vitra International AG: Rolf Fehlbaum
Design Zaha Hadid
Project Architect Patrik Schumacher
Consultant Architect Roland Mayer
Detail Design Patrik Schumacher, Signy Svalastoga
Design Team Simon Koumjian, Edgar González, Kar Wha Ho, Voon Yee-Wong, Craig Kiner, Cristina Verissimo, Maria Rossi, Daniel R. Oakley, Nicola Cousins, David Gomersall, Olaf Weishaupt
Models Daniel Chadwick, Tim Price
Project Manager, Construction Drawings and Building Supervision GPF & Assoziierte: Roland Mayer, Jürgen Roth, Shahriar Eetezadi, Eva Weber, Wolfgang Mehnert

Music-Video Pavilion [56–57]
Groningen, Netherlands, 1990
Client City Planning Department Groningen
Design Team Zaha Hadid with Graham Modlen, Urit Luden, Edgar González, Vincent Marol, Maria Rossi, Dianne Hunter-Gorman, Cristina Verissimo, Yousif Albustani, Craig Moffatt, Craig Kiner
Model Daniel Chadwick
Co-architect Karelse Van der Meer
Total Area 24.5 m² (4 levels: ground floor, 2 balconies and video room)

Hotel and Residential Complex [58]
Abu Dhabi, UAE, 1990
Client Sheikh Tahnoon bin Saeed Al Nahyan
Design Team Zaha Hadid with Vincent Marol, Craig Kiner, Yousif Albustani, Satoshi Ohashi, Patrik Schumacher, Daniel R. Oakley, Philippa Makin, Dianne Hunter-Gorman
Model Daniel Chadwick
Structural Engineer Ove Arup & Partners
Total Area 47,000 m² (2 retail floors, 1 office floor, 28 hotel floors)

Interzum 91 [59]
Gluzendorf, Germany, 1990
Exhibition Stand Design

London 2066 [60–61]
London, UK, 1991
Client British *Vogue*
Design Team Zaha Hadid with Daniel R. Oakley, Voon Yee-Wong, Graham Modlen, Craig Kiner, Cristina Verissimo, Yousif Albustani, Patrik Schumacher, Mascha Veech-Kosmatschof, Graeme Little
Computer Modelling Daniel R. Oakley

The Hague Villas [62]
The Hague, Netherlands, 1991
Design Team Zaha Hadid with Craig Kiner, Patrik Schumacher, Yousif Albustani, James Braam, Daniel R. Oakley, John Stuart, Cristina Verissimo, David Gomersall
Model Craig Kiner
Structural Engineer Ove Arup & Partners

The Great Utopia [63]
New York, USA, 1992
Design Team Zaha Hadid with Patrik Schumacher, Yousif Albustani, Daniel R. Oakley, David Gomersall, Simon Koumjian
Models Tim Price, Ademir Volic

Vision for Madrid [64]
Madrid, Spain, 1992
Design Team Zaha Hadid with Patrik Schumacher, Daniel R. Oakley, Simon Koumjian, Yousif Albustani, Craig Kiner, Paco Mejias

Arthotel Billie Strauss [65]
Nabern, Germany, 1992
Design Team Zaha Hadid with Patrik Schumacher, Yousif Albustani, Daniel R. Oakley, David Gomersall

Concert Hall [65]
Copenhagen, Denmark, 1992–1993
Design Zaha Hadid with Patrik Schumacher
Design Team Paul Brislin, Brian Ma Siy, John Comparelli, Nicola Cousins, Edgar González, Douglas Grieco, C. J. Lim, Mya Manakides, Guido Schwark
Structural Engineer Ove Arup & Partners
Acoustics Consultant Arup Acoustics: Malcolm Wright
Theatre Consultant Theatre Projects Consultants

Rheinauhafen Redevelopment [66–67]
Cologne, Germany, 1992
Design Zaha Hadid
Design Team Patrik Schumacher, Daniel R. Oakley, Craig Kiner, Yousif Albustani, Cathleen Chua, David Gomersall, John Stuart, Simon Koumjian
Model Tim Price

Carnuntum [68–69]
Vienna, Austria, 1993
Design Team Zaha Hadid with Patrik Schumacher, Douglas Grieco, Wendy Ing, Brian Ma Siy, Paola Sanguinetti, Edgar González, David Gomersall
Model Daniel Chadwick

Cardiff Bay Opera House [70–73]
Cardiff, UK, 1994–1996
Client Cardiff Bay Opera House Trust: The Rt Hon. Lord Crickhowell, Chairman
Design Zaha Hadid
Project Architect Brian Ma Siy
Design Team Patrik Schumacher, Ljiljana Blagojevic, Graham Modlen, Paul Brislin, Edgar González, Paul Karakusevic, David Gomersall, Tomás Amat Guarinos, Wendy Ing, Paola Sanguinetti, Nunu Luan, Douglas Grieco, Woody Yao, Voon Yee-Wong, Anne Save de Beaurecueil, Simon Koumjian, Bijan Ganjei, Nicola Cousins
Models Ademir Volic, Michael Kennedy, James Wink
Percy Thomas Partnership Ian Pepperell, Richard Roberts, Russell Baker, Richard Puckrin
Project Manager Stanhope Properties: Peter Rogers
Structural Engineer Ove Arup & Partners: Jane Wernick, David Glover, John Lovell
Services Consultant Ove Arup & Partners: Simon Hancock
Acoustics Consultant Arup Acoustics: Richard Cowell, Nigel Cogger
Theatre Consultant Theatre Projects Consultants: David Staples, Alan Russell, Anne Minors

Quantity Surveyor Gardiner & Theobald; Tillyard: Brett Butler, Peter Coxall
Arts Consultant AEA Consulting: Adrian Ellis, Jan Billington
Brief Consultant Inter Consult Culture: Charlotte Nassim
Construction Manager Lehrer McGovern Bovis: Alan Lansdell
Total Area 25,000 m²

Spittelau Viaducts [74–77]
Vienna, Austria, 1994–2005
Client SEG Developers
Design Zaha Hadid with Edgar González, Douglas Grieco, Paul Brislin, Patrik Schumacher, Woody Yao
Project Architects Woody Yao, Markus Dochantschi
Detail Design Zaha Hadid with Woody Yao, Markus Dochantschi, Wassim Halabi, Garin O'Aivazian, James Geiger
Design Team Clarissa Mathews, Paola Sanguinetti, Peter Ho, Anne Save de Beaurecueil, David Gomersall
Structural Engineer Friedreich & Partner
Total Area 2,600 m²

Blueprint Pavilion, Interbuild 95 [78]
Birmingham, UK, 1995
Clients Blueprint magazine; Montgomery Exhibitions Ltd
Design Zaha Hadid with Paul Brislin and Woody Yao
Design Team Tomás Amat Guarinos, Oliviero Godi, Maha Kutay, Clarissa Mathews, Graham Modlen, Anne Save de Beaurecueil, Leena Ibrahim
Computer Imagery Flexagon Studio: Thomas Quihano, Wassim Halabi
Structural Engineer Ove Arup & Partners: Rob Devey, Shiguru Hikone, Colin Jackson, Darren Sri-Tharan, Jane Wernick
Quantity Surveyor Tillyard: Brett Butler
Total Area 120 m²

42nd Street Hotel [79]
New York, USA, 1995
Clients Weiler Arnow Management Co; Milstein Properties
Design Team Zaha Hadid with Douglas Grieco, Peter Ho, Clarissa Mathews, Anne Save de Beaurecueil, Voon Yee-Wong, Woody Yao, Paul Brislin, Graham Modlen, Patrik Schumacher, David Gomersall, Bijan Ganjei
Model Richard Armiger
Images for Model Dick Stracker
Computer Imagery Rolando Kraeher
Structural Engineer Ove Arup & Partners
Total Area 180,000 m²

Spittalmarkt [80]
Berlin, Germany, 1995
Design Zaha Hadid with Patrik Schumacher
Competition Team Patrik Schumacher, Woody Yao, Wassim Halabi, David Gomersall, Graham Modlen
Design Development Patrik Schumacher, James Geiger

Lycée Français Charles de Gaulle [81]
London, UK, 1995
Design Team Zaha Hadid with Douglas Grieco, Edgar González, Paul Brislin, Brian Ma Siy, Paola Sanguinetti, Woody Yao, David Gomersall

Pancras Lane [81]
London, UK, 1996
Design Team Zaha Hadid with Brian Ma Siy, Paul Brislin, Edgar González, Patrik Schumacher, Douglas Grieco, Woody Yao, Paola Sanguinetti

Boilerhouse Extension [82]
London, UK, 1996
Client Victoria & Albert Museum
Design Team Zaha Hadid with Patrik Schumacher, Brian Ma Siy, Graham Modlen, Ljiljana Blagojevic, Paul Karakusevic, David Gomersall, Woody Yao, Markus Dochantschi, Wassim Halabi, Ivan Pajares Sanchez, Maha Kutay, Simon Yu, Tomás Amat Guarinos, James Geiger, Tilman Schall, Alan Houston
Structural Engineer Ove Arup & Partners: Jane Wernick
Building Services Ove Arup & Partners: Simon Hancock
Construction Manager Ove Arup & Partners (PMS): Peter Platt-Higgins
Cost Consultant Davis Langdon & Everest: Rob Smith
Total Area 10,000 m²

Wish Machine: World Invention [83]
Kunsthalle, Vienna, Austria, 1996
Client Kunsthalle Wien: Herbert Lachmeyer, Curator, Brigitte Felderer
Design Team Zaha Hadid with Patrik Schumacher, Simon Yu, Wassim Halabi, Markus Dochantschi, David Gomersall, Woody Yao, Paul Karakusevic
Total Area 900 m²

Master's Section, Venice Biennale [84]
Venice, Italy, 1996
Design Team Zaha Hadid with Patrik Schumacher, Markus Dochantschi, Woody Yao, Wassim Halabi, Garin O'Aivazian, David Gomersall, Simon Yu, Yousif Albustani, Giuseppe Anzalone Gherardi

Habitable Bridge [84–85]
London, UK, 1996
Client The Rt Hon. John Gummer, Secretary of State for the Environment; Royal Academy
Sponsor Thames Water
Design Team Zaha Hadid with Patrik Schumacher, Ljiljana Blagojevic, Paul Karakusevic, Graham Modlen, Woody Yao, Markus Dochantschi, Tilman Schall, Colin Harris, Thilo Fuchs, Shumon Basar, Katrin Kalden, Anne-Marie Foster
Models Alan Houston, Michael Howe
Computer Design Wassim Halabi, Simon Yu, Garin O'Aivazian
Structural Engineer Ove Arup & Partners: Jane Wernick, Sophie Le Bourva
Services Consultant Ove Arup & Partners: Simon Hancock, Dorte Rich Jorgensen
Transportation Consultant Ove Arup & Partners: John Shaw
Construction Manager Ove Arup & Partners: Harry Saradjian
Cost Consultant Davis Langdon & Everest: Rob Smith, Sam Mackenzie
Total Area 40,000 m²

La Fenice [86]
Venice, Italy, 1996
Design Team Zaha Hadid with Graham Modlen, Maha Kutay, Simon Yu
Computer Design Wassim Halabi

Philharmonic Hall [87]
Luxembourg, 1997
Client Ministry of Public Buildings
Design Zaha Hadid with Patrik Schumacher
Design Team Garin O'Aivazian, Markus Dochantschi, Woody Yao, Wassim Halabi, Jan Hübener, Anna Klingmann, Tilman Schall, Filipe Pereira, Shumon Basar, Mark Hemel, Yousif Albustani, Graham Modlen, Anuschka Kutz, David Gomersall
Total Area 7,100 m²

Landesgartenschau 1999 [88–91]
Weil am Rhein, Germany, 1996–1999
Client Landesgartenschau Weil am Rhein GmbH
Design Zaha Hadid with Patrik Schumacher and Mayer Bährle
Project Architect Markus Dochantschi
Project Team Oliver Domeisen, Wassim Halabi, Garin O'Aivazian, Barbara Pfenningstorff, James Lim
Models June Tamura, Jim Heverin, Jon Richards, Ademir Volic
Co-architect Mayer Bährle Freie Architekten BDA
Total Area 800 m²

Museum of Islamic Arts [92–93]
Doha, Qatar, 1997
Client State of Qatar
Design Zaha Hadid with Patrik Schumacher and Woody Yao
Design Team Shumon Basar, Graham Modlen, Markus Dochantschi, Anuschka Kutz, Garin O'Aivazian, Filipe Pereira, Ivan Pajares Sanchez, Wassim Halabi, Ali Mangera, Edgardo Torres, Julie Fisher, Andrew Schachman, Oliver Domeisen, Julie Richards, Irene Huttenrauch, Tia Lindgren
Total Area 28,000 m²

Sheikh Zayed Bridge [94–95]
Abu Dhabi, UAE, 1997–2010
Client Abu Dhabi Municipality
Design Zaha Hadid
Project Architect Graham Modlen
Project Team Garin O'Aivazian, Zahira Nazer, Christos Passas, Sara Klomps, Steve Power
Project Engineers Joe Barr, Mike King, Mike Davies; Highpoint Rendel
Structural Consultant Rendel Palmer & Tritton
Lighting Hollands Licht
Materials Piers, decking: reinforced concrete; arches: steel
Size 842 m (length) × 64 m (height) × 61 m (width)

Campus Center, Illinois Institute of Technology [96]
Chicago, Illinois, USA, 1998
Design Zaha Hadid with Patrik Schumacher
Design Team Yousif Albustani, Anuschka Kutz, Oliver Domeisen, Shumon Basar, Inken Witt, Jee-Eun Lee, Wassim Halabi, Ivan Pajares Sanchez, David Gomersall, Stéphane Hof, Woody Yao, Markus Dochantschi, Marco Guarnieri, Ali Mangera, Jim Heverin, Jon Richards, Terence Koh, Simon Yu, James Lim, Tilman Schall
Structural and Civil Engineer Ove Arup & Partners: Jane Wernick
Building Services Ove Arup & Partners: Simon Hancock

Acoustics Consultant Arup Acoustics: Andrew Nicol
Construction Manager Ove Arup & Partners: Peter Platt-Higgins
Quantity Surveyor Davis Langdon & Everest: Sam Mackenzie, Brian Irving
Fire and Means of Escape Ove Arup & Partners: Chris Barber
Information Technology Ove Arup & Partners: Volker Buscher
Urban Context Report Space Syntax Laboratory: Bill Hillier, Mark David Major

Lois & Richard Rosenthal Center for Contemporary Art [97–99]
Cincinnati, Ohio, USA, 1997–2003
Design Zaha Hadid
Project Architect Markus Dochantschi
Assistant Project Architect Ed Gaskin
Project Team Ana Sotrel, Jan Hübener, David Gerber, Christos Passas, Sonia Villaseca, James Lim, Jee-Eun Lee, Oliver Domeisen, Helmut Kinzler, Patrik Schumacher, Michael Wolfson, David Gomersall
Competition Team Shumon Basar, Oliver Domeisen, Jee-Eun Lee, Terence Koh, Marco Guarinieri, Stéphane Hof, Woody Yao, Ivan Pajares Sanchez, Wassim Halabi, Nan Atichapong, Graham Modlen
Study Models Chris Dopheide, Thomas Knüvener, Sara Klomps, Bergendy Cooke, Florian Migsch, Sandra Oppermann, Ademir Volic
Presentation Model Ademir Volic
Local Architect KZF Design: Donald L. Cornett, Mark Stedtefeld, Dale Beeler, Amy Hauck-Hamilton, Deb Lanius
Construction Manager Turner Construction Company: Craig Preston, Bill Huber, Bob Keppler
Structural Engineer THP Limited Inc: Shayne O. Manning, Murray Monroe, Andreas Greuel, Jason Jones
Acoustics Consultant Arup Acoustics: Neill Woodger, Andrew Nicol, Richard Cowell
Services Consultant Heapy Engineering: Ron Chapman, Gary Eodice, Kirby Morgan, Fred Grable
Security Consultant Steven R. Keller & Associates: Steven Keller, Pete Rondo
Theatre Consultant Charles Cosler Theatre Design
Lighting Consultant Office for Visual Interaction: Enrique Peiniger, Jean M. Sundin

The Mind Zone [100]
London, UK, 1998–2000
Client New Millennium Experience Company
Design Zaha Hadid
Project Architect Jim Heverin
Project Team Barbara Kuit, Jon Richards, Paul Butler, Ana Sotrel, Christos Passas, Graham Modlen, Oliver Domeisen
Competition Team Graham Modlen, Patrik Schumacher, Oliver Domeisen, Garin O'Aivazian, Simon Yu, Wassim Halabi, Jim Heverin, Jon Richards
Models Jon Richards, Jim Heverin, Eddie Can, Helmut Kinzler; A Models
Artist Liaison Doris Lockhart-Saatchi
Artists/Exhibit Collaborators Research Studios: Neville Brody; Richard Brown, Nancy Burson, Brian Butterworth, Helen Chadwick, Hussein Chalayan, Richard Deacon, Escape, Ryoji Ikeda, Herbert Lachmayer (with Matthias Fuchs and Sylvia Eckermann), Langlands & Bell, Ron Mueck, New Renaissance, Urs B. Roth, Gavin Turk
Structural Engineer Ove Arup & Partners
Building Services Ove Arup & Partners
Cladding Design Consultant DCAb
Construction Manager McAlpine/Laing Joint Venture
Principal Contractor Hypsos Expo
Steel Contractor Watson Steel Structures Ltd
GRP Contractor SP Offshore
Quantity Surveyor Davis Langdon & Everest
Lighting Consultant Hollands Licht

UNL/Holloway Road Bridge [101]
London, UK, 1998
Design Team Zaha Hadid with Christos Passas, Ali Mangera, Patrik Schumacher, Woody Yao, Sonia Villaseca, Eddie Can, Jorge Ortega, Helmut Kinzler
Structural Engineer Jane Wernick Associates
Building Services Ove Arup & Partners: Simon Hancock
Transport and Flow Capacity Ove Arup & Partners: Fiona Green
Construction Programme Ove Arup & Partners: Harry Saradjain
Quantity Surveyor Davis Langdon & Everest: James Woodrough

Pet Shop Boys World Tour [101]
1999
Design Zaha Hadid
Project Architect Oliver Domeisen
Design Team Bergendy Cooke, Jee-Eun Lee, Christos Passas, Caroline Voet, Susann Schweizer, Thomas Knüvener
Lighting Design Marc Brickman

Car Park and Terminus Hoenheim-Nord [102–105]
Strasbourg, France, 1998–2001
Client Compagnie des Transports Strasbourgeois
Design Zaha Hadid
Project Architect Stéphane Hof
Sketch Design Team Stéphane Hof, Sara Klomps, Woody Yao, Sonia Villaseca
Project Team Silvia Forlati, Patrik Schumacher, Markus Dochantschi, David Salazar, Caroline Voet, Eddie Can, Stanley Lau, David Gerber, Chris Dopheide, Edgar Gonzáles
Project Consultant Mayer Bährle Freie Architekten BDA: Roland Mayer
Local Architect Albert Grandadam
General Engineer Getas/Serue
Structural Engineer Luigi Martino
Total Area 25,000 m²; tram station: 3,000 m²

MAXXI: National Museum of XXI Century Arts [106–109]
Rome, Italy, 1998–2009
Client Italian Ministry of Culture
Design Zaha Hadid with Patrik Schumacher
Project Architect Gianluca Racana
Site Supervision Team Anja Simons, Paolo Matteuzzi, Mario Mattia
Design Team Anja Simons, Paolo Matteuzzi, Fabio Ceci, Mario Mattia, Maurizio Meossi, Paolo Zilli, Luca Peralta, Maria Velceva, Matteo Grimaldi, Amin Taha, Caroline Voet, Gianluca Ruggeri, Luca Segarelli
Competition Team Ali Mangera, Oliver Domeisen, Christos Passas, Sonia Villaseca, Jee-Eun Lee, James Lim, Julia Hansel, Sara Klomps, Shumon Basar, Bergendy Cooke, Jorge Ortega, Stéphane Hof, Marcus Dochantschi, Woody Yao, Graham Modlen, Jim Heverin, Barbara Kuit, Ana Sotrel, Hemendra Kothari, Zahira El Nazel, Florian Migsch, Kathy Wright, Jin Wananabe, Helmut Kinzler
Planning ABT s.r.l. (Rome)
Structure Anthony Hunt Associates (London); OK Design Group (Rome)
M&E Max Fordham & Partners (London); OK Design Group (Rome)
Lighting Equation Lighting Design (London)
Acoustics Paul Gillieron Acoustic (London)
Size 30,000 m²

Museum for the Royal Collection [110]
Madrid, Spain, 1999
Design Zaha Hadid with Patrik Schumacher
Design Team Sonia Villaseca, Caroline Voet, Jorge Ortega, Eddie Can, Paola Cattarin, Jee-Eun Lee, David Gomersall, Chris Dopheide, Silvia Forlati, J.R. Kim
Structural Engineer Ove Arup & Partners: David Glover, Ed Clark
Services Engineer Ove Arup & Partners: Simon Hancock
Cost Consultant Davis Langdon & Everest: Eloi Ruart
Museum Design Consultant Bruce McAllister

Reina Sofía Museum Extension [111]
Madrid, Spain, 1999
Design Zaha Hadid with Patrik Schumacher
Design Team Sonia Villaseca, Jorge Ortega, Eddie Can, Paola Cattarin, Christos Passas, Chris Dopheide, Bergendy Cooke, Jee-Eun Lee, Caroline Voet, Oliver Domeisen, David Gomersall, Electra Mikelides
Structural Engineer Ove Arup & Partners: David Glover, Ed Clark
Services Engineer Ove Arup & Partners: Simon Hancock
Cost Consultant Davis Langdon & Everest: Eloi Ruart
Museum Design Consultant Bruce McAllister

Rothschild Bank Headquarters and Furniture [111]
London, UK, 1999
Client N. M. Rothschild & Sons
Design Zaha Hadid
Design Team Graham Modlen, Barbara Kuit, Zahira Nazer, Oliver Domeisen
Models Florian Migsch, Bergendy Cooke, Thomas Knüvener, Jee-Eun Lee

Royal Palace Hotel and Casino [112]
Lugano, Switzerland, 1999
Design Zaha Hadid
Design Team Ali Mangera, Barbara Kuit, Thomas Knüvener, Paola Cattarin, Woody Yao, Patrik Schumacher, Jorge Ortega, Eddie Can, Silvia Forlati, Oliver Domeisen, Jee-Eun Lee, Bergendy Cooke
Collaborators Zahira Nazer, Jan Hübener, Yoash Oster
Structural Engineer Ove Arup & Partners: David Glover, Colin Jackson, Ed Clark
Environmental Engineer Ove Arup & Partners: Simon Hancock
Casino Consultant Edward Lyon Design

Metapolis, Charleroi/Danses [112]
Charleroi, Belgium, 1999
Design Zaha Hadid
Design Team Caroline Voet, Woody Yao, Stéphane Hof, Shumon Basar, Paola Cattarin, Bergendy Cooke, Chris Dopheide

Technical Consultant DCAb
Costume Consultant Susan Schweizer
Fabrics Consultant Marie O'Mahony
Couturier Thomas Zaepf

National Library of Quebec [113]
Montreal, Quebec, Canada, 2000
Design Zaha Hadid with Patrik Schumacher
Design Team Sonia Villaseca, Stéphane Hof, Chris Dopheide, Djordje Stojanovic, Dillon Lin, Lida Charsouli, Garin O'Aivazian, David Gerber, Andreas Durkin, Liam Young, Christos Passas, Sara Klomps
Competition Model Ademir Volic
Local Architect Albert Grandadam
Structural Engineer Adams Kara Taylor: Hanif Kara
Cost Consultants Davis Langdon & Everest: Guy Rezeau; Hanscomb Consultants
Environmental Engineer Max Fordham Partnership: Henry Luker, Sam Archer
Acoustics Consultants Arup Acoustics; Peutz & Associés France
Lighting Consultant Office for Visual Interaction: Enrique Peiniger, Jean M. Sundin

Bergisel Ski Jump [114–115]
Innsbruck, Austria, 1999–2002
Design Zaha Hadid
Project Manager Markus Dochantschi
Project Architect Jan Hübener
Project Team Matthias Frei, Cedric Libert, Silvia Forlati, Jim Heverin, Garin O'Aivazian, Sara Noel Costa de Araujo
Competition Team Ed Gaskin, Eddie Can, Yoash Oster, Stanley Lau, Janne Westermann
Structural Engineer Christian Aste
Project Management Georg Malojer
Services Engineers Technisches Büro Ing. Heinz Pürcher; Technisches Büro Matthias Schrempf; Peter Fiby
Ski Jump Technology Bauplanungsbüro Franz Fuchslueger
Lighting Consultant Office for Visual Interaction: Enrique Peiniger, Jean M. Sundin

Phæno Science Centre [116–121]
Wolfsburg, Germany, 2000–2005
Client Neulandgesellschaft GmbH, on behalf of the City of Wolfsburg
Design Zaha Hadid with Christos Passas
Co-architect Mayer Bährle Freie Architekten BDA
Project Architect Christos Passas
Assistant Project Architect Sara Klomps
Project Team Sara Klomps, Gernot Finselbach, David Salazar, Helmut Kinzler
Competition Team Christos Passas, Janne Westermann, Chris Dopheide, Stanley Lau, Eddie Can, Yoash Oster, Jan Hübener, Caroline Voet
Special Contributor Patrik Schumacher
Contributors Silvia Forlati, Günter Barczik, Lida Charsouli, Marcus Liermann, Kenneth Bostock, Enrico Kleinke, Constanze Stinnes, Liam Young, Chris Dopheide, Barbara Kuit, Niki Neerpasch, Markus Dochantschi
Project Architects (Mayer Bährle) Rene Keuter, Tim Oldenburg
Project Team (Mayer Bährle) Sylvia Chiarappa, Stefan Hoppe, Andreas Gaiser, Roman Bockemühl, Annette Finke, Stefanie Lippardt, Marcus Liermann, Jens Hecht, Christoph Volckmar
Structural Engineers Adams Kara Taylor; Tokarz Frerichs Leipold
Services Engineers NEK; Büro Happold
Cost Consultant Hanscomb GmbH
Lighting Consultants Fahlke & Dettmer; Office for Visual Interaction
Area 27,000 m²

Meshworks [122]
Rome, Italy, 2000
Design Zaha Hadid
Design Team Simon Koumjian, Patrik Schumacher, Caroline Voet
Installation Team Justin Porcano, Alfredo Greco, Michael Osmen, Daniel Arbelaez, Peter Kohn

Serpentine Gallery Pavilion [123]
London, UK, 2000
Client Serpentine Gallery
Design Zaha Hadid
Project Architect Jim Heverin
Project Manager Eric Gabriel
Structural Engineer CETEC Consultants
Quantity Surveyor Howard Associates
Lighting Consultant Maurice Brill Lighting Design
Contractor Gap Sails + Structures

British Pavilion, Venice Biennale [123]
Venice, Italy, 2000
Design Zaha Hadid
Project Architect Woody Yao
Design Team Eddie Can, Jan Hübener, Gianluca Racana
Installation Team Chris Dopheide, Alessandra Calglia, Justin Porcano, Michael Osmen, Daniel Arbelaez

Kunsthaus Graz [124]
Graz, Austria, 2000
Design Zaha Hadid with Patrik Schumacher
Design Team Sonia Villaseca, Stanley Lau, Paola Cattarin, David Gerber, Eddie Can, Gianluca Racana, Yoash Oster, Janne Westermann
Structural Engineer Adams Kara Taylor: Hanif Kara
Façade Consultant Adams Kara Taylor: Hanif Kara
Cost Consultant Davis Langdon & Everest: Sam Mackenzie

La Grande Mosquée de Strasbourg [125]
Strasbourg, France, 2000
Design Zaha Hadid with Patrik Schumacher
Design Team Ali Mangera, David Gerber, David Salazar, Jorge Ortega, Caroline Voet, Eddie Can, Patrik Schumacher, Woody Yao, Stéphane Hof, Hon Kong Chee, Steve Power, Edgar González, Garin O'Aivazian
Local Architect Albert Grandadam
Structural Engineer Adams Kara Taylor: Hanif Kara
Environmental Engineer Max Fordham Partnership: Henry Luker, Sam Archer
Acoustics Consultants Arup Acoustics; Peutz & Associés France
Lighting Consultant Office for Visual Interaction: Enrique Peiniger, Jean M. Sundin
Cost Consultant Davis Langdon & Everest: Guy Rezeau

Centro JVC Hotel [125]
Guadalajara, Mexico, 2000
Client Omnitrition de México
Design Zaha Hadid
Project Architect Jim Heverin
Project Team Helmut Kinzler, Edgar Gonzáles, Eddie Can, Jorge Ortega, Zulima Nieto, Jose Rojo
Structural Engineer Adams Kara Taylor
Building Services Büro Happold
Fire Consultant Arup Fire

Salerno Maritime Terminal [126–129]
Salerno, Italy, 2000–2016
Client Comune di Salerno Palazzo di Città
Design Zaha Hadid with Patrik Schumacher
Project Architect Paola Cattarin
Design Team Vincenzo Barilari, Andrea Parenti, Anja Simons, Giovanna Sylos Labini, Cedric Libert, Filippo Innocenti
Competition Team Paola Cattarin, Sonia Villaseca, Christos Passas, Chris Dopheide
Design Consultants:
Local Executive Architect Interplan Seconda: Alessandro Gubitosi
Costing Building Consulting: Pasquale Miele
Structural Engineers Ingeco: Francesco Sylos Labini; Ove Arup & Partners (preliminary design); Sophie Le Bourva
M&E Engineers Macchiaroli & Partners: Roberto Macchiaroli; Itaca: Felice Marotta; Ove Arup & Partners (preliminary design)
Maritime/Transport Engineer Ove Arup & Partners (London, UK): Greg Heigh
Lighting Equation Lighting Design (London, UK): Mark Hensmann
Main Contractor Passarelli SpA
Site Supervision:
Director of Works Gaetano Di Maio
Architecture Paola Cattarin
Structure Giampiero Martuscelli
MEP Roberto Macchiaroli
Contract Administration Pasquale Miele
Health & Safety Alessandro Gubitosi
Area 4,500 m²

Zaha Hadid Lounge [130]
Wolfsburg, Germany, 2001
Design Zaha Hadid
Project Architects Woody Yao, Djordje Stojanovic

Albertina Extension [130]
Vienna, Austria, 2001
Design Zaha Hadid with Patrik Schumacher
Project Architect Lars Teichmann
Design Team Ken Bostock, Dillon Lin, Tiago Correia, Sandra Oppermann, Raza Zahid

Temporary Museum, Guggenheim Tokyo [131]
Tokyo, Japan, 2001
Client Guggenheim Foundation
Design Zaha Hadid
Project Architect Patrik Schumacher
Design Team Gianluca Racana, Ken Bostock, Vivek Shankar
Total Area 7,000 m²

One-North Masterplan [131]
Singapore, 2002
Design Zaha Hadid with Patrik Schumacher
Project Director Markus Dochantschi
Project Architects (Masterplan Phase) David Gerber, Dillon Lin, Silvia Forlati
Project Team (Masterplan Phase) David Mah, Gunther Koppelhuber, Rodrigo O'Malley, Kim Thornton, Markus Dochantschi
Project Architects (Rochester Detail Planning Phase) Gunther Koppelhuber
Project Team (Rochester Detail Planning Phase) Kim Thornton, Hon Kong Chee, Yael Brosilovski, Fernando Pérez Vera
Competition Team David Gerber, Edgar González, Chris Dopheide, David Salazar, Tiago Correia, Ken Bostock, Patrik Schumacher, Paola Cattarin, Dillon Lin, Barbara Kuit, Woody Yao
Models Riann Steenkamp, Chris Dopheide, Ellen Haywood, Helena Feldman
Presentation Models Delicatessen Design Ltd: Ademir Volic
Urban Strategy Lawrence Barth, Architectural Association
Infrastructural Engineer Ove Arup & Partners: Simon Hancock, Ian Carradice, David Johnston
Infrastructural Audits JTC Consultants Pte Ltd
Transport Engineer MVA: Paul Williams, Tim Booth
Landscape Architect Cicada Pte Ltd
Lighting Consultant Lighting Planners Associates: Kaoru Mende
Planning Tool Consultant B Consultants: Tom Barker, Graeme Jennings

BMW Plant Central Building [132–135]
Leipzig, Germany, 2001–2005
Design Zaha Hadid with Patrik Schumacher
Project Architects Jim Heverin, Lars Teichmann
Project Team Matthias Frei, Jan Hübener, Annette Bresinsky, Manuela Gatto, Fabian Hecker, Cornelius Schlotthauer, Wolfgang Sunder, Anneka Wegener, Markus Planteu, Robert Neumayr, Christina Beaumont, Achim Gergen, Caroline Anderson
Competition Team Lars Teichmann, Eva Pfannes, Ken Bostock, Stéphane Hof, Djordje Stojanovic, Leyre Villoria, Liam Young, Christiane Fashek, Manuela Gatto, Tina Gregoric, Cesare Griffa, Yasha Jacob Grobman, Filippo Innocenti, Zetta Kotsioni, Debora Laub, Sarah Manning, Maurizio Meossi, Robert Sedlak, Niki Neerpasch, Eric Tong, Tiago Correia
Partner Architects IFB Dr Braschel AG; Anthony Hunt Associates
Structural Engineers IFB Dr Braschel AG; WPW Ingenieure GmbH
M&E Engineer IFB Dr Braschel AG
Cost Consultant IFB Dr Braschel AG
Lighting Design Equation Lighting Design
Landscape Architect Gross.Max

Ordrupgaard Museum Extension [136–139]
Copenhagen, Denmark, 2001–2005
Design Zaha Hadid
Project Architect Ken Bostock
Design Team Caroline Krogh Andersen
Competition Team Patrik Schumacher, Ken Bostock, Adriano De Gioannis, Sara Noel Costa de Araujo, Lars Teichmann, Vivek Shankar, Cedric Libert, Tiago Correia
Model Riann Steenkamp
Associate Architect PLH Arkitekter

Structural Engineers Jane Wernick Associates; Birch & Krogboe
M&E Design Ove Arup & Partners; Birch & Krogboe
Lighting Consultant Arup Lighting
Acoustics Consultant Birch & Krogboe

Maggie's Centre Fife [140–141]
Kirkcaldy, UK, 2001–2006
Client Maggie's Centre
Design Zaha Hadid
Project Architects Jim Heverin, Tiago Correia
Project Team Zaha Hadid, Jim Heverin, Tiago Correia
Structural Engineer Jane Wernick Associates
Services Engineer K.J. Tait Engineers
Underground Drainage SKM Anthony Hunts
Quantity Surveyor CBA Chartered Quantity Surveyors
Planning Supervisor Reiach & Hall Architects
Landscape Architect Gross.Max
Building Surveyor GLM Ltd
Total Area 250 m²

López de Heredia Pavilion [142–143]
Haro la Rioja, Spain, 2001–2006
Client López de Heredia
Design Zaha Hadid
Project Architect Jim Heverin
Project Team Tiago Correia, Matthias Frei, Ana M. Cajiao
Partner Architect IOA Arquitectura: Joan Ramon Rius, Nuria Ayala, Xavier Medina, Candi Casadevall
Structural Engineer Jane Wernick Associates
M&E Engineer Ove Arup & Partners: Ann Dalzell

Monographic Exhibition [144]
Rome, Italy, 2002
Design Zaha Hadid with Patrik Schumacher
Project Architects Gianluca Racana, Woody Yao
Design Team Tiago Correia, Adriano De Gioannis, Barbara Pfenningstorff, Ana M. Cajiao, Maurizio Meossi, Manuela Gatto, Thomas Vietzke, Natalie Rosenberg, Ken Bostock, Barbara Kuit, Christos Passas, Sara Klomps
Local Architect ABT s.r.l.

Price Tower Arts Center [144]
Bartlesville, Oklahoma, USA, 2002
Design Zaha Hadid with Patrik Schumacher
Project Architect Markus Dochantschi
Project Team Matias Musacchio, Ana M. Cajiao, Jorge Seperizza, Mirco Becker, Tamar Jacobs, Viggo Haremst, Christian Ludwig, Ed Gaskin

City of Towers, Venice Biennale [145]
Venice, Italy, 2002
Client Alessi
Design Zaha Hadid with Patrik Schumacher
Design Team Thomas Vietzke, Natalie Rosenberg, Woody Yao

Pierres Vives [146–148]
Montpellier, France, 2002–2012
Client Département de l'Hérault
Design Zaha Hadid
Project Architect Stéphane Hof
Project Team Joris Pauwels, Philipp Vogt, Rafael Portillo, Jaime Serra, Renata Dantas, Melissa Fukumoto, Jens Borstelmann, Kane Yanegawa, Loreto Flores, Edgar Payan, Lisamarie Villegas Ambia, Karouko Ogawa, Stella Nikolakaki, Hon Kong Chee, Caroline Andersen, Judith Reitz, Olivier Ottevaere, Achim Gergen, Daniel Baerlecken, Yosuke Hayano, Martin Henn, Rafael Schmidt, Daniel Gospodinov, Kia Larsdotter, Jasmina Malanovic, Ahmad Sukkar, Ghita Skalli, Elena Perez, Andrea B. Castè, Lisa Cholmondeley, Douglas Chew, Larissa Henke, Steven Hatzellis, Jesse Chima, Adriano De Gioannis, Simon Kim, Stéphane Carnuccini, Samer Chamoun, Ram Ahronov, Ross Langdon, Ivan Valdez, Yacira Blanco, Marta Rodriguez, Leonardo Garcia, Sevil Yazici, Hussam Chakouf, Marie-Perrine Placais, Monica Noguero, Naomi Fritz, Stephanie Chaltiel
Local Architect Blue Tango (design phase); Chabanne et Partenaires (execution phase)
Structure Ove Arup & Partners
Services Ove Arup & Partners (concept design); GEC Ingénierie
Acoustics Rouch Acoustique: Nicolas Albaric
Cost Consultant Gec-LR: Ivica Knezovic
Size 35,000 m²

Museum Brandhorst [149]
Munich, Germany, 2002
Design Zaha Hadid with Patrik Schumacher
Project Architect Barbara Pfenningstorff
Design Team Adriano De Gioannis, Cornelius Schlotthauer, Maurizio Meossi, Filippo Innocenti, Rocio Paz, Eric Tong, Ana M. Cajiao, Flavio La Gioia, Viggo Haremst, Manuela Gatto, Tamar Jacobs, Thomas Vietzke, Natalie Rosenberg, Christos Passas
Structural Engineer ASTE: Andreas Glatzl, Christian Aste
NME and General Lighting Max Fordham Partnership: Henry Luker
Exterior Lighting Office for Visual Interaction: Enrique Peiniger, Jean M. Sundin
Glass Façade RFR Ingénieurs: Jean-François Blassel

Zorrozaurre Masterplan [150–151]
Bilbao, Spain, 2003–
Client Management Committee for the Urban Development of the Peninsula of Zorrozaurre, Bilbao
Design Zaha Hadid with Patrik Schumacher
Project Architects Manuela Gatto, Fabian Hecker
Project Team Juan Ignacio Aranguren, Daniel Baerlecken, Yael Brosilovski, Helen Floate, Marc Fornes, James Gayed, Steven Hatzellis, Alvin Huang, Yang Jingwen, Gunther Koppelhuber, Graham Modlen, Brigitta Lenz, Susanne Lettau, Fernando Pérez Vera, Judith Reitz, Marta Rodriguez, Jonathan Smith, Marcela Spadaro, Kim Thornton, Zhi Wang
Local Architect Arkitektura Eta Hiriginza Bulegoa S.A.
Engineer Ove Arup & Partners
Landscape Architect Gross.Max
Total Area 60 hectares

BBC Music Centre [152]
London, UK, 2003
Client BBC
Design Zaha Hadid with Patrik Schumacher
Project Architects Steven Hatzellis, Graham Modlen, Ergian Alberg
Project Team Karim Muallem, Ram Ahronov, Adriano De Gioannis, Simon Kim, Yansong Ma, Rafael Schmidt, Markus Planteu
Structural Engineer Ove Arup & Partners: Bob Lang
Services Consultant Ove Arup & Partners: Nigel Tonks
Acoustics Consultant Arup Acoustics: Richard Cowell

Theatre Consultant Anne Minors Performance Consultants
Cost Consultant Davis Langdon & Everest: Sam Mackenzie

Desire [152]
Graz, Austria, 2003
Design Zaha Hadid with Patrik Schumacher
Project Architect Rocio Paz
Design Team Filippo Innocenti, Zetta Kotsioni, Alexander de Looz
Structural Engineer B Consultants

Guggenheim Museum Taichung [153]
Taichung, China, 2003
Design Zaha Hadid with Patrik Schumacher
Project Architect Dillon Lin
Design Team Jens Borstelmann, Thomas Vietzke, Yosuke Hayano
Production Team Adriano De Gioannis, Selim Mimita, Juan Ignacio Aranguren, Ken Bostock, Elena Perez, Ergian Alberg, Rocio Paz, Markus Planteu, Simon Kim
Structural Engineer Adams Kara Taylor: Hanif Kara, Andrew Murray, Sebastian Khourain, Reuben Brambleby, Stefano Strazzullo
Services Consultant IDOM Bilbao
Cost Consultant IDOM UK Ltd; IDOM Bilbao
Model Photography David Grandorge

The Snow Show [154]
Lapland, Finland, 2003–2004
Design Zaha Hadid with Patrik Schumacher
Project Architects Rocio Paz, Woody Yao
Design Team Yael Brosilovski, Thomas Vietzke, Helmut Kinzler
Structural Engineer Adams Kara Taylor
Lighting Design Zumtobel Illuminazione s.r.l., with HFG-Karlsruhe Scenography class, Prof. M. Simon

NYC 2012 Olympic Village [154]
New York, USA, 2004
Design Zaha Hadid with Patrik Schumacher
Project Manager Studio MDA: Markus Dochantschi
Project Architect Tiago Correia
Design Team Ana M. Cajiao, Daniel Baerlecken, Judith Reitz, Simon Kim, Dillon Lin, Yosuke Hayano, Ergian Alberg, Yael Brosilovski, Daniel Li, Yang Jingwen, Li Zou, Laura Aquili, Jens Borstelmann, Juan Ignacio Aranguren
Urban Strategy Lawrence Barth
Structural Engineer Ove Arup & Partners: Bob Lang
Infrastructural Engineer Ove Arup & Partners: Ian Carradice
Transport Engineer Ove Arup & Partners: David Johnston
Building Services Ove Arup & Partners: Andrew Sedgwick
Lift Consultant Ove Arup & Partners: Mike Summers, Warrick Gorrie
Mechanical Engineer Ove Arup & Partners: Emmanuelle Danisi
Fire Consultant Ove Arup & Partners: Barbara Lane, Tony Lovell
Security Consultant Arup Security Consulting: John Haddon, Simon Brimble
Façade Consultant Ove Arup & Partners: Edith Mueller
Materials Ove Arup & Partners: Clare Perkins
Lighting Design L'Observatoire International: Hervé Descottes
Landscape Architect Gross.Max: Eelco Hooftmann

Cost Consultant Davis Langdon Adamson: Nick Butcher, Ethan T. Burrows
Video Animation and 3D Visuals Neutral
Photography David Grandorge

Hotel Puerta America – Hoteles Silken [155]
Madrid, Spain, 2003–2005
Client Grupo Urvasco
Design Zaha Hadid
Project Architect Woody Yao
Project Design Thomas Vietzke, Yael Brosilovski, Patrik Schumacher
Design Team Ken Bostock, Mirco Becker
Total Area 1,200 m²

Guangzhou Opera House [156–161]
Guangzhou, China, 2003–2010
Client Guangzhou Municipal Government
Design Zaha Hadid
Project Directors Woody Yao, Patrik Schumacher
Project Leader Simon Yu
Project Team Jason Guo, Yang Jingwen, Long Jiang, Ta-Kang Hsu, Yi-Ching Liu, Zhi Wang, Christine Chow, Cyril Shing, Filippo Innocenti, Lourdes Sánchez, Hinki Kwong, Junkai Jian
Competition Team (1st Stage) Filippo Innocenti, Matias Musacchio, Jenny Huang, Hon Kong Chee, Markus Planteu, Paola Cattarin, Tamar Jacobs, Yael Brosilovski, Viggo Haremst, Christian Ludwig, Christina Beaumont, Lorenzo Grifantini, Flavio La Gioia, Nina Safainia, Fernando Pérez Vera, Martin Henn, Achim Gergen, Graham Modlen, Imran Mahmood
Competition Team (2nd Stage) Cyril Shing, YanSong Ma, Yosuke Hayano, Adriano De Gioannis, Barbara Pfenningstorff
Local Design Institute Guangzhou Pearl River Foreign Investment Architectural Designing Institute
Structural Engineers SHTK; Guangzhou Pearl River Foreign Investment Architectural Designing Institute
Façade Engineers KGE Engineering
Building Services Guangzhou Pearl River Foreign Investment Architectural Designing Institute
Acoustics Consultant Marshall Day Acoustics
Theatre Consultant ENFI
Lighting Design Consultant Beijing Light & View
Project Management Guangzhou Municipal Construction Group Co., Ltd
Construction Management Guangzhou Construction Engineering Supervision Co., Ltd
Structure, Services and Acoustics (Competition Stage) Ove Arup & Partners
Cost Consultant Guangzhou Jiancheng Engineering Costing Consultant Office Ltd
Principal Contractor China Construction Third Engineering Bureau Co., Ltd
Area 70,000 m²

High-Speed Train Station Napoli-Afragola [162–163]
Naples, Italy, 2003–2017
Client TAV s.p.a.
Design Zaha Hadid with Patrik Schumacher
Project Director Filippo Innocenti
Project Architects Paola Cattarin, Roberto Vangeli
Design Team Michele Salvi, Federico Bistolfi, Cesare Griffa, Paolo Zilli, Mario Mattia, Tobias Hegemann, Chiara Baccarini, Alessandra Bellia, Serena Pietrantonj, Roberto Cavallaro, Karim Muallem, Luciano Letteriello, Domenico Di Francesco, Marco Guardinccerri, Davide Del Giudice

Competition Team Fernando Pérez Vera, Ergian Alberg, Hon Kong Chee, Cesare Griffa, Karim Muallem, Steven Hatzellis, Thomas Vietzke, Jens Borstelmann, Robert Neumayr, Elena Perez, Adriano De Gioannis, Simon Kim, Selim Mimita
Structural and Geotechnical Engineers Adams Kara Taylor: Hanif Kara, Paul Scott; Interprogetti: Giampiero Martuscelli
Environmental Engineers Max Fordham Partnership: Henry Luker, Neil Smith; Studio Reale: Francesco Reale, Vittorio Criscuolo Gaito
Cost Consultant Building Consulting: Pasquale Miele
Fire Safety Macchiaroli & Partners: Roberto Macchiaroli
Transport Engineer JMP: Max Matteis
Landscape Architect Gross.Max: Eelco Hooftman
Acoustics Consultant Paul Gillieron Acoustic
Building Regulation, Coordination Local Team Interplan Seconda: Alessandro Gubitosi
General Contractor Ati Astaldi
Total Area 20,000 m²

California Residence [164]
San Diego, California, USA, 2003–
Design Zaha Hadid with Patrik Schumacher
Associate Kenneth Bostock
Project Architects Claudia Wulf, Elke Presser
Design Team (previous proposals) Christos Passas, Barbara Pfenningsdorf, Daniel Fiser, Tyen Masten, Marcela Spadaro, Theodora Ntatsopouloui
Civil Engineering/Planning Florez Engineering, Inc
Contractor Wardell Builders
Construction Morley
Electrical Design Nutter
Estimator Gilbane Building Company
Façade Consultant Front, Inc
Lighting Office for Visual Interaction
Local Architect Public I Architecture + Planning
Soils Consultant Geocon, Inc
Structural and Systems Engineers Nabih Youssef Associates
Sustainable Design Consultant KEMA Services, Inc
Vertical Landscape Consultant Patrick le Blanc
Site Area 20,000 m²
Floor Area 11,000 m²

Eleftheria Square Redesign [165]
Nicosia, Cyprus, 2005–2017
Client The City of Nicosia
Design Zaha Hadid with Patrik Schumacher
Project Architect Christos Passas
Project Team Dimitris Akritopoulos, Anna Papachristoforou, Thomas Frings, Irene Guerra, Jesus Garate, Ivan Ucros, Javier Ernesto-Lebie, Vincent Nowak, Mathew Richardson, Giorgos Mailis, Sevil Yazici, Ta-Kang Hsu, Sylvia Georgiadou, Phivos Scroumbelos, Marilena Sophocleous
Competition Project Architects Christos Passas, Saffet Kaya Bekiroglu
Competition Team Inanc Eray, Viviana Muscettola, Michelle Pasca, Daniel Fiser
Structural Engineer Hyperstatic Engineering
M&E/Services Unemec
Lighting Consultant Kardorff Ingenieure
Project Management Modinos & Vrahimis
Cost Consultant MDA Consulting Ltd
Open Area 35,300 m²
Interior Area 7,175 m²

Nordpark Cable Railway [166–169]
Innsbruck, Austria, 2004–2007
Client Innsbrucker Nordkettenbahnen GmbH
Design Zaha Hadid with Patrik Schumacher
Project Architect Thomas Vietzke
Design Team Jens Borstelmann, Markus Planteu
Production Team Caroline Krogh Andersen, Makakrai Suthadarat, Marcela Spadaro, Anneka Wegener, Adriano De Gioannis, Peter Pichler, Susann Berggren
Local Partner Office/Building Management Malojer Baumanagement
Structural Engineer Bollinger Grohmann Schneider Ziviltechniker
Façade Planning Pagitz Metalltechnik GmbH
Contractor Strabag AG
Engines and Cables Contractor Leitner GmbH
Planning Advisors ILF Beratende Ingenieure ZT GmbH; Malojer Baumanagement
Concrete Base Baumann + Obholzer ZT
Bridge Engineer ILF Beratende Ingenieure ZT GmbH
Lighting Zumtobel Illuminazione s.r.l.

Riverside Museum [170–171]
Glasgow, UK, 2004–2011
Client Glasgow City Council
Design Zaha Hadid Architects
Project Director Jim Heverin
Project Architect Johannes Hoffman
Project Team Achim Gergen, Agnes Koltay, Alasdair Graham, Andreas Helgesson, Andy Summers, Aris Giorgiadis, Brandon Buck, Christina Beaumont, Chun Chiu, Daniel Baerlecken, Des Fagan, Electra Mikelides, Elke Presser, Gemma Douglas, Hinki Kwong, Jieun Lee, Johannes Hoffmann, Laymon Thaung, Lole Mate, Malca Mizrahi, Markus Planteu, Matthias Frei, Mikel Bennett, Ming Cheong, Naomi Fritz, Rebecca Haines-Gadd, Thomas Hale, Tyen Masten
Competition Team Malca Mizrahi, Michele Pasca di Magliano, Viviana R. Muscettola, Mariana Ibanez, Larissa Henke
Services Büro Happold
Acoustics Büro Happold
Fire Safety FEDRA
Cost Consultants Capita Symonds
Project Management Capita Symonds
Area 11,300 m²

City Life Milano [172–175]
Milan, Italy, 2004–2017
Client City Life Consortium
Design Zaha Hadid with Patrik Schumacher
Project Director Gianluca Racana
Residential Complex:
Project Architect Maurizio Meossi
Design Team Vincenzo Barilari, Cristina Capanna, Giacomo Sanna, Paola Bettinsoli, Gianluca Bilotta, Fabio Ceci, Veronica Erspamer, Arianna Francioni, Stefano Iacopini, Mario Mattia, Serena Pietrantonj, Florindo Ricciuti, Giulia Scaglietta, Giovanna Sylos Labini, Anja Simons, Marta Suarez, Tamara Tancorre, Giuseppe Vultaggio, Massimiliano Piccinini, Samuele Sordi, Alessandra Belia
Site Supervision Team Cristina Capanna, Veronica Erspamer, Stefano Iacopini, Giulia Scaglietta, Florindo Ricciuti
Competition Team Simon Kim, Yael Brosilovski, Adriano De Gioannis, Graham Modlen, Karim Muallem, Daniel Li, Yang Jingwen, Tiago Correia, Ana Cajiao, Daniel Baerlecken, Judith Reitz
Structural Consultant MSC Associati
M&E Consultant Hilson Moran Italia
Fire Safety Ing. Silvestre Mistretta
Project Specifications Building Consulting
General Contractor City Contractor s.r.l.
Electrical Consultant (Construction) Impes
Mechanical Engineering Consultant (Construction) Panzeri
Façade Planning Permasteelisa Group
Size 7 buildings from 5–13 storeys; 38,000 m² gross surface, 230 units; 2-storey undergound car park 50,000 m²
Generali Tower/Retail Space:
Project Architect Paolo Zilli
Design Team Andrea Balducci Caste, H. Goswin Rothenthal, Gianluca Barone, Carles S. Martinez, Arianna Russo, Giuseppe Morando, Pierandrea Angius, Vincenzo Barilari, Stefano Paiocchi, Sara Criscenti, Alexandra Fisher, Agata Banaszek, Marco Amoroso, Alvin Triestanto, Letizia Simoni, Subharthi Guha, Marina Martinez, Luis Miguel Samanez, Santiago F. Achury, Massimo Napoleoni, Massimiliano Piccinini, Annarita Papeschi, Martha Read, Peter McCarthy, Line Rahbek, Mario Mattia, Matteo Pierotti, Shahd Abdelmoneim
Site Supervision Team Andrea Balducci Caste, Pierandrea Angius, Vincenzo Barilari
Competition Team Simon Kim, Yael Brosilovski, Adriano De Gioannis, Graham Modlen, Karim Muallem, Daniel Li, Yang Jingwen, Tiago Correia, Ana Cajiao, Daniel Baerlecken, Judith Reitz
Management J&A/Ramboll
Structural Engineer AKT; Redesco (tower); Holzner&Bertagnolli + Cap (basement)
M&E Max Fordham + Manens-TIFS
Project Specifications Building Consulting
Façade Planning Arup
Lifts Jappsen
Fire Safety Mistretta
Height 170 m; 44 storeys
Total Gross Floor Area 122,700 m²: office 67,000 m²; retail 15,000 m²; parking, storage & plantrooms 40,700 m²

Zaragoza Bridge Pavilion [176–179]
Zaragoza, Spain, 2005–2008
Client Expoagua Zaragoza 2008
Design Zaha Hadid with Patrik Schumacher
Project Architect Manuela Gatto
Project Team Fabian Hecker, Matthias Baer, Soohyun Chang, Feng Chen, Atrey Chhaya, Ignacio Choliz, Federico Dunkelberg, Dipal Kothari, Maria José Mendoza, José M. Monfa, Marta Rodriguez, Diego Rosales, Guillermo Ruiz, Lucio Santos, Hala Sheikh, Marcela Spadaro, Anat Stern, Jay Suthadarat
Competition Team Feng Chen, Atrey Chhaya, Dipal Kothari
Engineering Consultant Ove Arup & Partners
Cost Consultant Ove Arup & Partners; IDOM
Size 270 m (length); 185 m from island to right bank, plus 85 m from island to Expo riverbank

London Aquatics Centre [180–185]
London, UK, 2005–2011
Client Olympic Delivery Authority
Design Zaha Hadid Architects
Project Director Jim Heverin
Project Architects Glenn Moorley, Sara Klomps
Project Team Alex Bilton, Alex Marcoulides, Barbara Bochnak, Carlos Garijo, Clay Shorthall, Ertu Erbay, George King, Giorgia Cannici, Hannes Schafelner, Hee Seung Lee, Kasia Townend, Nannette Jackowski, Nicholas Gdalewitch, Seth Handley, Thomas Soo, Tom Locke, Torsten Broeder, Tristan Job, Yamac Korfali, Yeena Yoon
Project Team (Competition) Saffet Kaya Bekiroglu (project architect), Agnes Koltay, Feng Chen, Gemma Douglas, Kakakrai Suthadarat, Karim Muallem, Marco Vanucci, Mariana Ibanez, Sujit Nair
Sports Architects S&P Architects
Structural Engineer Ove Arup & Partners
Services Ove Arup & Partners
Fire Safety Arup Fire
Acoustics Arup Acoustics
Façade Engineers Robert-Jan Van Santen Associates
Lighting Design Arup Lighting
Kitchen Design Winton Nightingale
Maintenance Access Reef
Temporary Construction Edwin Shirley Staging
Security Consultant Arup Security
AV & IT Consultants Mark Johnson Consultants
Access Consultants Access=Design
CDM Coordinator Total CDM Solutions
BREEAM Consultant Southfacing
Quantity Surveyor & Project Manager CLM
Main Contractor Balfour Beatty
Timber Sub-Contractor Finnforest Merk GmbH
Concrete Sub-Contractor Morrisroe
Site Area 36,875 m²
Total Floor Area:
Legacy Basement 3,725 m²; ground floor 15,137 m²; first floor 10,168 m²
Olympic Basement 3,725 m²; ground floor 15,402 m²; first floor 16,387 m²; seating area 7,352 m²
Footprint Area 15,950 m² (legacy); 21,897 m² (Olympic)

Capital Hill Residence [186–187]
Moscow, Russia, 2006–
Design Zaha Hadid with Patrik Schumacher
Project Architect Helmut Kinzler
Project Designer Daniel Fiser
Design Team Anat Stern, Thomas Sonder, Muthahar Khan, Kristina Simkeviciute, Talenia Phua Gajardo, Mariana Ibanez, Marco Vanucci, Lourdes Sánchez, Ebru Simsek, Daniel Santos
Initial Design Stage Leader Tetsuya Yamazaki
Project Manager Capital Group: Natalia Savich
Local Architect Mar Mimarlik Ltd
Structural Engineer ENKA Engineering
M&E Engineer ENKA Engineering
Electrical Engineer HB Electric
Façade Engineer Group 5F
AV Sound Ideas UK Ltd
BMS/Security Siemens
Lighting Design Light Tecnica
Interior Design Candy & Candy Ltd
Pool Design Rainbow Pools Ltd
Landscape Panorama Landscape Design
Water Feature Design Invent Water Features Ltd
General Contractor ENKA Construction & Industry Co, Inc
Façade Consultant Wagner Group

CMA CGM Headquarters Tower [188–189]
Marseille, France, 2006–2011
Client CMA CGM
Design Zaha Hadid with Patrik Schumacher
Project Director Jim Heverin

Project Architect Stéphane Vallotton
Project Team Karim Muallem, Simone Contasta, Leonie Heinrich, Alvin Triestanto, Muriel Boselli, Eugene Leung, Bhushan Mantri, Jerome Michel, Nerea Feliz, Prashanth Sridharan, Birgit Eistert, Evelyn Gono, Marian Ripoll, Andres Flores, Pedja Pantovic
Competition Team Jim Heverin, Simon Kim, Michele Pasca di Magliano, Viviana R. Muscettola
Local Architect SRA; RTA
Local Engineers Coplan (Provence)
Structural Engineer Ove Arup & Partners
Services Ove Arup & Partners
Façade Engineers Ove Arup & Partners; Robert-Jan Van Santen Associates
Cost Consultant R2M
Engineers Ove Arup & Partners
Façade Consultant Robert-Jan Van Santen Associates
Acoustician Albarique Rouche
Structural Contractor GTM Sud
Façade Contractor Epitekh
Lift Contractor Schindler
Paint Contractor SLVR
Floor finishes Unimarbres; Mattout
Waterproofiing SMAC
Mechanical G+S
Electrical Cegelec Sud-Est; Santerne
Ironmongery Chiri
Wall finishes Menuiserie Lazer (glass); MBA/Staff Parisien (plaster)
Doors Mayvaert/Portafeu
Area 94,000 m²

Deutsche Guggenheim [190]
Berlin, Germany, 2005
Client Deutsche Bank
Design Zaha Hadid with Patrik Schumacher
Lead Designer Helmut Kinzler
Design Team Tetsuya Yamazaki, Yael Brosilovski, Saleem Abdel-Jalil, Joris Pauwels, Manuela Gatto, Fabian Hecker, Gernot Finselbach, Judith Reitz, Daniel Baerlecken, Setsuko Nakamura

Urban Nebula [190]
London, UK, 2007
Client London Design Festival 2007
Design Zaha Hadid with Patrik Schumacher
Project Team Charles Walker, Daniel Dendra
Structural Engineer Adams Kara Taylor
Manufacturer Aggregate Industries
Size 240 cm × 1,130 cm × 460 cm

Lilas [191]
London, UK, 2007
Client Serpentine Gallery
Design Zaha Hadid with Patrik Schumacher
Project Architect Kevin McClellan
Structural Engineer Ove Arup & Partners
Steel Fabrication Sheetfabs Ltd
Membrane Fabrication Base Structures Ltd
Lighting Design Zumtobel Illuminazione s.r.l.
Furniture Established & Sons, Kenny Schachter, Sawaya & Moroni, Serralunga, Max Protetch, Swarovski
Total Area 310 m²; 5.5 m (height) × 22.5 m (width) × 22.5 m (length)

Investcorp Building [192–193]
Oxford, UK, 2006–2015
Client Middle East Centre, St Antony's College
Design Zaha Hadid
Project Director Jim Heverin
Project Associate Johannes Hoffmann, Ken Bostock
Project Architect Alex Bilton
Project Team Sara Klomps, Goswin Rothenthal, Andy Summers, George King, Luke Bowler, Barbara Bochnak, Yeena Yoon, Saleem A. Jalil, Theodora Ntatsopoulou, Mireira Sala Font, Amita Kulkarni
Structural Engineer AKTII
Mechanical/Electrical/Acoustic Engineer Max Fordham
Client's M&E Consultant Elementa
Façade Supplier Frener + Reifer
Façade Consultant Arup Façade Engineering
Client's Façade Consultant Eckersley O'Callaghan
Contractor BAM
Project Manager Bidwells
Lighting Design Arup Lighting
Cost Consultant Sense Cost Ltd
Fire Engineer Arup Fire
Planning Supervision JPPC Oxford
Forestry and Arboriculture Consultant Sarah Venner
Access David Bonnet
Landscape Design Gross.Max
CDM Andrew Goddard Associates
Visualization Cityscape
Overall Site Area 1,580 m²
Gross Internal Area 1,127 m²
Building Footprint 700 m²

Kartal-Pendik Masterplan [194–195]
Istanbul, Turkey, 2006–
Client Greater Istanbul Municipality and Kartal Urban Regeneration Association
Design Zaha Hadid with Patrik Schumacher
Overall Project Architect Bozana Komljenovic
Stage 2 Project Team Amit Gupta, Marie-Perrine Placais, Susanne Lettau, Elif Erdine, Jimena Araiza
Stage 1 Project Leaders Bozana Komljenovic, DaeWha Kang
Stage 1 Project Team Sevil Yazici, Vigneswaran Ramaraju, Brian Dale, Jordan Darnell, Oznur Erboga
Competition Leaders DaeWha Kang, Saffet Kaya Bekiroglu
Competition Team Sevil Yazici, Daniel Widrig, Elif Erdine, Melike Altinisik, Ceyhun Baskin, Inanc Eray, Fulvio Wirz, Gonzalo Carbajo
Total Area 5.5 million m² (555 hectares)

Dubai Opera House [196–197]
Dubai, UAE, 2006
Design Zaha Hadid with Patrik Schumacher
Project Director Charles Walker
Project Architect Nils-Peter Fischer
Project Team Melike Altinisik, Alexia Anastasopoulou, Dylan Baker-Rice, Domen Bergoc, Shajay Bhooshan, Monika Bilska, Alex Bilton, Elizabeth Bishop, Torsten Broeder, Cristiano Ceccato, Alessio Constantino, Mario Coppola, Brian Dale, Ana Valeria Emiliano, Elif Erdine, Camilla Galli, Brandon Gehrke, Aris Giorgiadis, Pia Habekost, Michael Hill, Shao-Wei Huang, Chikara Inamura, Alexander Janowsky, DaeWha Kang, Tariq Khayyat, Maren Klasing, Britta Knobel, Martin Krcha, Effie Kuan, Mariagrazia Lanza, Tyen Masten, Jwalant Mahadevwala, Rashiq Muhamadali, Mónica Noguero, Diogo Brito Pereira, Rafael Portillo, Michael Powers, Rolando Rodriguez-Leal, Federico Rossi, Mireia Sala Font, Elke Scheier, Rooshad Shroff, William Tan, Michal Treder, Daniel Widrig, Fulvio Wirz, Susu Xu, Ting Ting Zhang
Project Director (Competition) Graham Modlen
Project Architect (Competition) Dillon Lin
Competition Team Christine Chow, Daniel Dendra, Yi-Ching Liu, Simone Fuchs, Larissa Henke, Tyen Masten, Lourdes Sánchez, Johannes Schafelner, Swati Sharma, Hooman Talebi, Komal Talreja, Claudia Wulf, Simon Yu
Engineering Consultant Ove Arup & Partners: Steve Roberts
Acoustics Consultant Arup Acoustics: Neill Woodger
Theatre Consultant Anne Minors Performance Consultants
Lighting Consultant Office for Visual Interaction

Evelyn Grace Academy [198–199]
Brixton, London, UK, 2006–2010
Client School trust: ARK Education Government: DCSF
Design Zaha Hadid with Patrik Schumacher
Project Director Lars Teichmann
Project Architect Matthew Hardcastle
Project Team Lars Teichmann, Matthew Hardcastle, Bidisha Sinha, Henning Hansen, Lisamarie Villegas Ambia, Enrico Kleinke, Judith Wahle, Christine Chow, Guy Taylor, Patrick Bedarf, Sang Hilliges, Hoda Nobakhti
Project Manager Capita Symonds
Engineers Arup
Quantity Surveyors Davis Langdon
Landscape Gross.Max
Acoustic Consultants Sandy Brown Associates
Fire Engineer SAFE
CDM Co-ordinator Arup
Size 10,745 m²

Dongdaemun Design Plaza [200–205]
Seoul, South Korea, 2007–2014
Design Zaha Hadid with Patrik Schumacher
Project Leader Eddie Can Chiu-Fai
Project Manager Craig Kiner, Charles Walker
Project Team Kaloyan Erevinov, Martin Self, Hooman Talebi, Carlos S. Martinez, Camiel Weijenberg, Florian Goscheff, Maaike Hawinkels, Aditya Chandra, Andy Chang, Arianna Russo, Ayat Fadaifard, Josias Hamid, Shuojiong Zhang, Natalie Koerner, Jae Yoon Lee, Federico Rossi, John Klein, Chikara Inamura, Alan Lu
Competition Team Kaloyan Erevinov, Paloma Gormley, Hee Seung Lee, Kelly Lee, Andres Madrid, Deniz Manisali, Kevin McClellan, Claus Voigtmann, Maurits Fennis
Structural Engineer ARUP
MEPF Services Engineer ARUP
Lighting Consultant ARUP
Acoustics Consultant ARUP
Landscape Architect Gross.Max
Façade Consultant Group 5F
Geometry Consultant Evolute
Quantity Surveyor Davis Langdon & Everest
Local Architects Samoo Architects
Local Consultants Structure Postech
Mechanical Samoo Mechanical Consulting (SMC)
Electrical and Telecom Samoo TEC
Façade M&C
Civil Saegil Engineering & Consulting
Landscape Dong Sim Won
Fire Korean Fire Protection

Regium Waterfront [206]
Reggio, Italy, 2007–
Design Zaha Hadid with Patrik Schumacher
Project Architect Filippo Innocenti
Design Team Michele Salvi, Roberto Vangeli, Andrea Balducci Caste, Luciano Letteriello, Fabio Forconi, Giuseppe Morando, Johannes Weikert, Deepti Zachariah, Gonzalo Carbajo
Structural Engineer Adams Kara Taylor: Hanif Kara
M&E Engineer Max Fordham Partnership: Neil Smith
Cost Surveyor Building Consulting: Alba De Pascale, Edoardo Limea
Maritime Structures Studio Prima: Pietro Chiavaccini, Maurizio Verzoni

582–606 Collins Street [207]
Melbourne, Australia, 2015–
Client Landream, Australia
Design Zaha Hadid with Patrik Schumacher
Director Gianluca Racana
Project Director Michele Pasca di Magliano
Project Architect Juan Camilo Mogollon
Project Team Johannes Elias, Hee Seung Lee, Cristina Capanna, Sam Mcheileh, Luca Ruggeri, Nhan Vo, Michael Rogers, Gaganjit Singh, Julia Hyoun Hee Na, Massimo Napoleoni, Ashwanth Govindaraji, Maria Tsironi, Kostantinos Psomas, Marius Cernica, Veronica Erspamer, Cyril Manyara, Megan Burke, Ahmed Hassan, Effie Kuan
Local Architect PLUS Architecture
Structural Engineering Robert Bird Group
Building Services Engineering and Sustainability ADP Consulting
Planning Consultant URBIS
Quantity Surveyor WT Partnership
Façade Consultant AURECON
Landscape Designer OCULUS
Wind Engineering MEL Consultants
Traffic Engineer RATIO
Building Surveyor PLP
Fire Engineer OMNII
Waste Management Leigh Design
Pedestrian Modelling ARUP
Acoustics Acoustic Logic
Land Surveyor Bosco Jonson
Visualizations VA
Gross Floor Area 70,075 m²

Messner Mountain Museum Corones [208–209]
South Tyrol, Italy, 2013–2015
Client Skirama Kronplatz/Plan de Corones
Design Zaha Hadid with Patrik Schumacher
Project Architect Cornelius Schlotthauer
Design Team Cornelius Schlotthauer, Peter Irmscher
Execution Team Peter Irmscher, Markus Planteu, Claudia Wulf
Structural Engineer IPM
Mechanical Engineer & Fire Protection Jud & Partner
Electrical Engineer Studio GM
Lighting Design Zumtobel
Principal Contractors Kargruber und Stoll (concrete); Pichler Stahlbau (façade); B&T Bau & Technologie (façade panels)
Gross Floor Area 1,000 m²
Elevation 2,275 m

D'Leedon [210]
Singapore, 2007–2014
Client CapitaLand-led consortium
Design Zaha Hadid with Patrik Schumacher
Project Directors Michele Pasca di Magliano, Vivivana Muscettola
Project Team Ludovico Lombardi, Clara Martins, Loreto Flores, Stephan Bohne, Amita Kulkarni, Soomeen Hahm, Yung-Chieh Huang Kanop Mangklapruk, Marina Martinez, Andres Moroni, Juan Camilo Mogollon, Michael Rissbacher, Luca Ruggeri, Luis Miguel Samanez, Nupur Shah, Puja Shah, Muhammed Shameel, Shankara Subramaniam, Manya Uppal, Katrina Wong, Paolo Zilli, Andrea Balducci Caste, Kutbuddin Nadiadi, Effie Kuan, Helen Lee, Hee Seung Lee, Annarita Papeschi, Feng Lin, Bianca Cheung, Dominiki Dadatsi, Kelly Lee, Jeonghoon Lee, Hoda Nobakhti, Judith Wahle, Zhong Tian, Akhil Laddha, Naomi Chen, Jee Seon Lim, Line Rahbek, Hala Sheikh, Carlos S. Martinez, Arianna Russo, Peter McCarthy, Sevil Yazici, Sandra Riess, Federico Rossi, Eleni Pavlidou, Federico Dunkelberg, Evan Erlebacher, Gorka Blas, Bozana Komljenovic, Sophie Le Bienvenu, Jose M. Monfa, Selahattin Tuysuz, Edward Calver
Concept Team Michele Pasca di Magliano, Viviana Muscettola, Ta-Kang Hsu, Emily Chang, Helen Lee, Kelly Lee
Local Architect RSP
Structural Engineering AECOM
M&E Engineering (Concept) Max Fordham & Partners
M&E Engineering BECA
Quantity Surveyor Gardiner & Theobald, London
Landscape Architect Gross.Max (concept); ICN
Lighting Design LPA
Acoustics Engineering Aciron
Total Floor Area 204,387 m² (over 7 towers); 150 m (height)

King Abdullah Financial District Metro Station [211]
Riyadh, Saudi Arabia, 2012–
Client Arriyadh Development Authority
Design Zaha Hadid with Patrik Schumacher
Project Principal Gianluca Racana
Project Director Filippo Innocenti
Project Associate Fulvio Wirz
Project Architect Gian Luca Barone
Project Team Abdel Halim Chehab, Alexandros Kallegias, Alexandre Kuroda, Arian Hakimi Nejad, Carine Posner, David Wolthers, Domenico Di Francesco, Izis Salvador Pinto, Jamie Mann, Manuele Gaioni, Marco Amoroso, Mario Mattia, Massimo Napoleoni, Mohammadali Mirzaei, Niki Okala, Nima Shoja, Roberto Vangeli, Sohith Perera, Stefano Iacopini
Competition Team Alexandre Kuroda, Fei Wang, Lisa Kinnerud, Jorge Mendez-Caceres
Structural Engineer Buro Happold
Services Buro Happold
Transport and Civil Engineering Buro Happold
Fire Engineering Buro Happold
Façade Consultant NewTecnic
Cost Consultant AECOM
Total Area 20,434 m²

Mobile Art Pavilion for Chanel [212–215]
Hong Kong, Tokyo, New York, Paris, 2007–2008
Client Chanel
Design Zaha Hadid with Patrik Schumacher
Project Architects Thomas Vietzke, Jens Borstelmann
Project Team Helen Lee, Claudia Wulf, Erhan Patat, Tetsuya Yamazaki, Daniel Fiser
Structural Engineer Ove Arup & Partners
Cost Consultant Davis Langdon & Everest
Main Contractor/Tour Operator ESS Staging
FRP Manufacturing Stage One Creative Services Ltd
Materials Façade cladding: fibre-reinforced plastic; roof: PVC; ETFE roof lights; secondary structure: aluminium extrusions; primary structure: 74 tons of steel (pavilion: 69 tons; ticket office: 5 tons); 1,752 different steel connections
Total Area 700 m²

Heydar Aliyev Centre [216–217]
Baku, Azerbaijan, 2007–2012
Client The Republic of Azerbaijan
Design Zaha Hadid and Patrik Schumacher with Saffet Kaya Bekiroglu
Project Architect Saffet Kaya Bekiroglu
Project Team Sara Sheikh Akbari, Shiqi Li, Phil Soo Kim, Marc Boles, Yelda Gin, Liat Muller, Deniz Manisali, Lillie Liu, Jose Lemos, Simone Fuchs, Jose Ramon Tramoyeres, Yu Du, Tahmina Parvin, Erhan Patat, Fadi Mansour, Jaime Bartolomé, Josef Glas, Michael Grau, Deepti Zachariah, Ceyhun Baskin, Daniel Widrig, with special thanks to Charles Walker
Site Area 111,292 m²
Auditorium Capacity 1,000 people
Unique Glassfibre Reinforced Polyester Panels 13,000 panels (40,000 m²)
Glassfibre Reinforced Concrete Panels 3,150 GRC panels (10,000 m²)
Glassfibre Reinforced Polyester Panel Production Max 70 unique panels per day
Museum Space Frame 12,569 members / 3,266 nodes
Auditorium Space Frame 17,269 members / 4,513 nodes
Total Surface Space Frame Area Approx. 33,000 m²
Internal Skin Fixing Plates 70,000
Surface Area of Inner Skin Over 22,000 m²
Metal Deck Roof Purlins at Tender Stage 3,607 – total length 10,092 m
Metal Roof Deck Trays at Tender Stage 3,936 – total area 26,853 m²
Roof Area 39,000 m²
Main Contractor and Architect of Record DiA Holding
Structural Engineer Tuncel Engineering AKT
Mechanical Engineer GMD Project
Electrical Consultant HB Engineering
Façade Consultant Werner Sobek
Fire Etik Fire Consultancy
Acoustics Mezzo Stüdyo
Geotechnical Engineer Enar Engineering
Infrastructure Sigal
Lighting MBLD

'Opus' Tower [218–219]
Dubai, UAE, 2007–
Client Omniyat Properties
Design Zaha Hadid with Patrik Schumacher
Project Director Christos Passas
Competition Team Christos Passas (Project Architect), Daniel Baerlecken, Gemma Douglas, Alvin Huang, Paul Peyrer-Heimstaett, Saleem Al-Jalil
Base Built Design Team Vincent Nowak (Project Architect/Development Phase), Dimitris Akritopoulos, Chiara Ferrari, Thomas Frings, Jesus Garate, Sylvia Georgiadou, Javier Ernesto-Lebie, Wenyuan Peng, Paul Peyrer-Heimstaett, Phivos Skroumbelos, Marilena Sophocleous
Supervision Team Fabian Hecker (Project Associate),

Barbara Bochnak (Lead Architect), Dimitris Kolonis (Senior Architect), Tomasz Starczewski, Kwanphil Cho, Bruno Pereira
Hotel Interior Design Team Reza Esmaeeli (Project Architect), Eider Fernandez Eibar (Design Coordination), Laura Micalizzi (Senior Interior Designer), Emily Rohrer (Senior Interior Designer), Stella Nikolakaki, Alexandra Fischer, Raul Forsoni, Chrysi Fradellou, Sofia Papageorgiou, Christos Sazos, Kwanphil Cho, Andri Shalou, Eleni Mente (Senior Landscape Designer)
Luxury Apartments Design Team Spyridon Kaprinis (Project Architect), Thomas Frings, Chrysi Fradellou, Sofia Papageorgiou, Carlos Luna
Consultants (Design Stage):
Project Management Gleeds (London)
Local Architects Arex Consultants (Dubai)
Structural Engineer Whitbybird (London)
Fire Engineering Safe (London)
Lift Consultants Roger Preston Dynamics (London)
Traffic Consultants Cansult Limited (Dubai)
Contractors Nasa / Multiplex (main contractor), Permasteelisa (façade contractor)
Consultants (Construction Stage):
Local Architects BSBG (Dubai)
Structural Engineer BG&E (Dubai)
Fire Engineering Design Confidence (Dubai)
Façade Consultants Koltay Facades (Dubai)
Lighting Consultants DPA (Dubai/UK)
Acoustic Consultants PMK (Dubai)
Interior Consultants HBA (Dubai)
Traffic Consultants Al Tourath (Dubai)
Lift Consultants Lerch Bates (Dubai)
Security Consultants Control Risks (Dubai)
Kitchen Consultants MCTS (Dubai)
Quantity Surveying Consultants HQS (Dubai)
AV Consultants EntireTech (Dubai)
Contractors Brookfield Multiplex (main contractor), Alu Nasa (façade contractor)

Jockey Club Innovation Tower [220–223]
Hong Kong, China, 2007–2014
Client Hong Kong Polytechnic University
Design Zaha Hadid with Patrik Schumacher
Project Director Woody Yao
Project Leader Simon K. M. Yu
Project Team Hinki Kong, Jinqi Huang, Bianca Cheung, Charles Kwan, Juan Liu, Junkai Jian, Zhenjiang Guo, Uli Blum, Long Jiang, Yang Jingwen, Bessie Tam, Koren Sin, Xu Hui, Tian Zhong
Competition Team Hinki Kwong, Melodie Leung, Long Jiang, Zhenjiang Guo, Yang Jingwen, Miron Mutyaba, Pavlos Xanthopoulus, Margarita Yordanova Valova
Local Architects AGC Design Ltd; AD+RG
Structural and Geotechnical Engineers Ove Arup & Partners; Hong Kong Ltd
Building Services Ove Arup & Partners; Hong Kong Ltd
Landscape Team 73 Hong Kong Ltd
Acoustics Consultant Westwood Hong & Associates Ltd
Total Area 15,000 m²; 78 m (height)

Dominion Office Building [224–225]
Moscow, Russia, 2012–2015
Client Peresvet Group/Dominion-M Ltd
Design Zaha Hadid with Patrik Schumacher
Design Director Christos Passas
Project Architects Veronika Ilinskaya, Kwanphil Cho
Interior Design Team Emily Rohrer, Raul Forsoni, Veronika Ilinskaya, Kwanphil Cho
Contributors Hussam Chakouf, Reza Esmaeeli, Thomas Frings
Art Installation Bruno Pereira
Concept Design Design Director: Christos Passas; Project Architect: Yevgeniy Beylkin; Design Team: Juan Ignacio Aranguren C, Yevgeniy Beylkin, Simon Kim, Agnes Koltay, Larisa Henke, Tetsuya Yamazaki
Local Architect AB Elis Ltd
Façade Consultant Ove Arup (London, UK)
Structural Engineer Mosproject
Concrete Engineer PSK Stroiltel Promstroicontract
MEP & General Contractor Stroigroup
Electrical MEP Novie Energiticheskie Reshenia
Façade Contractors StroyBit, Prostie reshenia/ ALUCOBOND™
Glazing Contractor MBK Stroi
Interior Contractor LCC Contractcity
GRC Architectura Blagopoluchie/Facade Light
Lighting Consultant FisTechenergo
Furniture Contractor M Factor
Total Floor Area 21,184 m²
Footprint 62 m × 50.5 m

Eli & Edythe Broad Art Museum [226–227]
East Lansing, Michigan, USA, 2007–2012
Client Michigan State University
Design Zaha Hadid with Patrik Schumacher
Project Director Craig Kiner
Project Architect Alberto Barba
Project Team Ruven Aybar, Michael Hargens, Edgar Payan Pacheco, Sophia Razzaque, Arturo Revilla, Charles Walker
Project Director (Competition) Nils-Peter Fischer
Project Architects (Competition) Britta Knobel, Fulvio Wirz
Competition Team Daniel Widrig, Melike Altinisik, Mariagrazia Lanza, Rojia Forouhar
Structural Consultant Adams Kara Taylor: Hanif Kara
Environmental/M&E Consultant Max Fordham Partnership: Henry Luker

Danjiang Bridge [228–231]
Taipei, China, 2015–
Client Directorate General of Highways, Taiwan, China
Status Design stage
Design Zaha Hadid with Patrik Schumacher
Competition Project Directors Charles Walker, Manuela Gatto
Competition Project Architect Shao-wei Huang
Competition Design Associate Paulo Flores
Competition Lead Designer Saman Saffarian
Competition Project Team Evgeniya Yatsyuk, Paul Bart, Sam Sharpe, Silviya Barzakova, Julian Lin, Ramon Weber
Delivery Project Director Cristiano Ceccato
Delivery Project Architect Shao-wei Huang
Delivery Project Team Carlos Michel-Medina, Chien-shuo Pai, Julian Lin
Delivery Project BIM Support Paul Ehret
Lead Structural Engineering Consultancy and JV Partner Leonhardt, Andrä und Partner, Germany
Local Engineering Consultants and JV Partner Sinotech Engineering Consultants
Lighting Designer Chroma33 Architectural Lighting Design, Taiwan, China
Length 920 m
Height of Supporting Mast 175 m

Abu Dhabi Performing Arts Centre [232–233]
Abu Dhabi, UAE, 2008–
Client Tourism Development & Investment Company of Abu Dhabi
Design Zaha Hadid with Patrik Schumacher
Project Director Nils-Peter Fischer
Project Architects Britta Knobel, Daniel Widrig
Project Team Jeandonne Schijlen, Melike Altinisik, Arnoldo Rabago, Zhi Wang, Rojia Forouhar, Jaime Serra Avila, Diego Rosales, Erhan Patat, Samer Chamoun, Philipp Vogt, Rafael Portillo
Structural, Fire, Traffic and Building Services Consultants WSP Group, with WSP (Middle East): Bill Price, Ron Slade
Acoustics Consultant Sound Space Design: Bob Essert
Façade Sample Construction King Glass Engineering Group
Theatre Consultant Anne Minors Performance Consultants
Cost Consultant Gardiner & Theobald: Gary Faulkner
Total Area 62,770 m²

Bee'ah Headquarters [234–237]
Sharjah, UAE, 2014–
Client Bee'ah
Status Design stage
Design Zaha Hadid with Patrik Schumacher
Director Charles Walker
Project Director Tariq Khayyat
Project Architect (Construction Phase) Sara Sheikh Akbari
Project Team John Simpson, Gerry Cruz, Drew Merkle, Maria Chaparro, Matthew Le Grice
ZHA Project Architect (Design Phase) Kutbuddin Nadiadi
Project Team (Design Phase) Gerry Cruz, Drew Merkle, Vivian Pashiali, Matthew Le Grice, Alia Zayani, Alessandra Lazzoni, Dennis Brezina, Yuxi Fu, Xiaosheng Li, Edward Luckmann, Eleni Mente, Kwanphil Cho, Mu Ren, Harry Ibbs, Mostafa El Sayed, Suryansh Chandra, Thomas Jensen, Alexandra Fischer, Spyridon Kaprinis, John Randle, Bechara Malkoun, Reda Kessanti, Eider Fernandez-Eibar, Carolina López-Blanco, Matthew Johnston, Sabrina Sayed, Zohra Rougab, Carl Khourey, Anas Younes, Lauren Barclay, Mubarak Al Fahim
Structure/Façade Buro Happold (London)
MEP Atelier Ten (London)
Cost Gardiner & Theobald (London)
Landscape Francis Landscape (Beirut)
Local Architect Bin Dalmouk
Renders MIR
Site Area 90,000 m²
Floor Area 7,000 m²
Height 18 m

Port House [238–241]
Antwerp, Belgium, 2009–2016
Client Port of Antwerp
Design Zaha Hadid with Patrik Schumacher
Project Director Joris Pauwels
Project Architect Jinmi Lee
Project Team Florian Goscheff, Monica Noguero, Kristof Crolla, Naomi Fritz, Sandra Riess, Muriel Boselli, Susanne Lettau
Competition Team Kristof Crolla, Sebastien Delagrange, Paulo Flores, Jimena Araiza, Sofia Daniilidou, Andres Schenker, Evan Erlebacher, Lulu Aldihani

Executive Architect and Cost Consultant Bureau Bouwtechniek
Structural Engineers Studieburo Mouton Bvba
Services Engineers Ingenium Nv
Acoustic Engineers Daidalos Peutz
Restoration Consultant Origin
Fire Protection FPC
Principal Contractors Interbuild, Groven+ (façade), Victor Buyck Steel Construction (steel)
Total Floor Area 20,800 m² (12,800 m² above grade)
Site Area 16,400 m²

Library & Learning Centre [242–245]
Vienna, Austria, 2008–2013
Client University of Economics Vienna
Design Zaha Hadid with Patrik Schumacher
Project Architect Cornelius Schlotthauer
Project Team Construction Enrico Kleinke, Markus Planteu, Vincenco Cocomero, Peter Irmscher, Katharina Jacobi, Constanze Stinnes, Peter Hornung, Frédéric Beaupère, Mirjam Matthiessen, Marc-Philipp Nieberg, Tom Finke, Kristoph Nowak, Susanne Lettau, Jahann Shah Beyzavi, Florian Goscheff, Daniela Nenadic, Judith Wahle, Rassul Wassa, Julian Breinersdorfer, Nastasja Schlaf, Muhammed Patat, Elisabeth Dirnbacher
Structural Engineers Vasko und Partner Ingenieure
M&E Engineers Vasko und Partner Ingenieure
Façade Engineers ARUP Deutschland GmbH
Lighting Engineers Arup Berlin
Fire Protection HHP West, Bielefeld
Site Supervision Ingenos Gobiet ZT GmbH, IC Consulenten Ziviltechniker GmbH
Project Team Competition Cornelius Schlotthauer, Marc-Philipp Nieberg, Enrico Kleinke, Kristoph Nowak, Stefan Rinnebach, Romy Heiland, Richard Baumgartner
Area 28,000 m²

Burnham Pavilion [246]
Chicago, Illinois, USA, 2009
Client Burnham Plan Centennial
Design Zaha Hadid with Patrik Schumacher
Project Architects Jens Borstelmann, Thomas Vietzke
Project Team Teoman Ayas, Evan Erlebacher
Local Architect Thomas Roszak
Structural Engineers Rockey Structures
Fabricator Fabric Images
Lighting & Electrical Tracey Dear
Film Installation Thomas Gray, The Gray Circle
Sound Design Lou Mallozzi, Experimental Sound Studio
Area 120 m²

Wangjing Soho [247]
Beijing, China, 2009–2014
Client Soho China Ltd
Design Zaha Hadid with Patrik Schumacher
Project Director Satoshi Ohashi
Project Associate Armando Solano
Project Team Yang Jingwen, Christoph Klemmt, Shu Hashimoto, Yung-Chieh Huang, Rita Lee, Samson Lee, Feng Lin, Seungho Yeo, Di Ding, Xuexin Duan, Chaoxiong Huang, Ed Gaskin, Bianca Cheung, Chao-Ching Wang, John Klein, Ho-Ping Hsia, Yu Du, Sally Harris, Oliver Malm, Rashiq Muhamadali, Matthew Richardson
Competition Team Satoshi Ohashi, Christiano Ceccato, Inanc Eray, Ceyhun Baskin, Chikara Inamura, Michael Grau, Hoda Nobakhati, Yevgeniya Pozigun, Michal Treder

Structural Consultant Adams Kara Taylor UK (competition), CCDI Beijing (SD), (DD), (CD)
Façade Consultant Arup Facade HK (SD), Inhabitat Beijing (DD)
MEP, VT, Fire Safety, Sustainability Consultant Hoare Lea UK (competition), Arup Engineers (SD)
Site Area 115,393 m²
Gross Floor Area 521,265 m² (392,265 m² above grade, 129,000 m² below grade)
Footprint Area 21,000 m²

King Abdullah Petroleum Studies & Research Center [248–249]
Riyadh, Saudi Arabia, 2009–2016
Client Saudi Aramco
Design Zaha Hadid with Patrik Schumacher
Project Directors Lars Teichmann, Charles Walker
Design Director DaeWha Kang
Project Leaders Fabian Hecker (research centre); Michael Powers (conference centre); Brian Dale, Henning Hansen (library); Fulvio Wirz (musalla/IT centre); Elizabeth Bishop (façades/2D documentation); Saleem A. Jalil, Maria Rodero (masterplan); Lisamarie Ambia, Judith Wahle (interiors); Bozana Komljenovic (2D documentation); John Randle (specifications); John Szlachta (3D documentation coordinator)
Construction Leads John Simpson (site associate), Alejandro Diaz Fernandez, Brian Dale (interiors), Elizabeth Bishop, Michal Wojtkiewicz (façades and canopies), Monika Bilska, Malgorzata Kowalczyk (services coordination), Henning Hansen, Ayca Vural Cutts, Michael Powers (structure), Ayca Vural Cutts, Sara Criscenti (external landscape)
Project Team Adrian Krezlik, Alexander Palacio, Amdad Chowdhury, Amit Gupta, Anas Younes, Andres Arias Madrid, Annarita Papeschi, Aritz Moriones, Ayca Vural Cutts, Britta Knobel, Camiel Weijenberg, Carine Posner, Claire Cahill, Claudia Dorner, DaChun Lin, Daniel Fiser, Daniel Toumine, David Doody, David Seeland, Deniz Manisali, Elizabeth Keenan, Evan Erlebacher, Fernanda Mugnaini, Garin O'Aivazian, Giorgio Radojkovic, Inês Fontoura, Jaimie-Lee Haggerty, Javier Rueda, Jeremy Tymms, Julian Jones, Jwalant Mahadevwala, Lauren Barclay, Lauren Mishkind, Malgorzata Kowalczyk, Mariagrazia Lanza, Melike Altinisik, Michael Grau, Michael McNamara, Michal Wojtkiewicz, Mimi Halova, Mohammad Ali Mirzaei, Mohammed Reshdan, Monika Bilska, Muriel Boselli, MyungHo Lee, Nahed Jawad, Natacha Viveiros, Navvab Taylor, Neil Vyas, Nicola McConnell, Pedro Sanchez, Prashanth Sridharan, Roxana Rakhshani, Saahil Parikh, Sara Criscenti, Sara Saleh, Seda Zirek, Shaju Nanukuttan, Shaun Farrell, Sophie Davison, Sophie Le Bienvenu, Stefan Brabetz, Stella Dourtme, Steve Rea, Suryansh Chandra, Talenia Phua Gajardo, Theodor Wender, Yu Du
Competition Design Team Lisamarie Ambia, Monika Bilska, Martin Krcha, Maren Klasing, Kelly Lee, Hannes Schafelner, Judith Schafelner, Ebru Simsek, Judith Wahle, Hee Seung Lee, Clara Martins, Anat Stern, Daniel Fiser, Thomas Sonder, Kristina Simkeviciute, Talenia Phua Gajardo, Erhan Patat, Dawna Houchin, Jwalant Mahadevwala
Engineering Arup
Interior Woods Bagot
Landscape Gross.Max
Lighting Office for Visual Interaction
Catering and Kitchen Eastern Quay and GWP
Exhibition Design Event
Branding & Signage Elmwood and Bright 3D

Library Consulting Tribal
Cost/Design Project Management Davis Langdon
Area 66,000 m²

Galaxy Soho [250–251]
Beijing, China, 2009–2012
Client Soho China Ltd
Design Zaha Hadid with Patrik Schumacher
Project Director Satoshi Ohashi
Associate Cristiano Ceccato
Project Architect Yoshi Uchiyama
Project Team Kelly Lee, Rita Lee, Eugene Leung, Lillie Liu, Rolando Rodriguez-Leal, Seung-ho Yeo. DD Phase: Dorian Bybee, Michael Grau, Shu Hashimoto, Shao-Wei Huang, Chikara Inamura, Lydia Kim, Christoph Klemmt, Yasuko Kobayashi, Raymond Lau, Wang Lin, Yereem Park, Tao Wen, Stephan Wurster. SD Phase: Samer Chamoun, Michael Hill, Tom Wuenschmann, Shuojiong Zhang
Competition Team DaeWha Kang (Lead Designer), Monika Bilska, Elizabeth Bishop, Diogo Brito, Brian Dale, Kent Gould, Jwalant Mahadevwala, Michael Powers, Vignesh Ramaraju
Local Design BIAD Beijing Institute of Architectural Design
Lighting Lightdesign
Contractor China Construction First Division Group Construction & Development Co., Ltd
Plot Area 46,965 m²
Total Floor Area 332,857 m²
Above Ground 4 towers, 15 floors (12 office floors and 3 retail floors)
Max. Height 67 m
Materials, Exterior 3 mm aluminium exterior cladding, insulated glass, stone
Materials, Interiors Glass, terrazzo, GRG, tile stainless steel, gypsum board painted
Structure Concrete Construction (8.4 m spans)
Floor-to-Floor Heights Retail floors 5.4 m, office floors 3.5 m
Landscape Stone, glass, stainless steel
Exterior Furniture Stone, FRP

Sky Soho [252–255]
Shanghai, China, 2010–2014
Client Soho China Ltd
Design Zaha Hadid with Patrik Schumacher
Project Director Manuela Gatto
Associate Consultant Satoshi Ohashi
Project Architects Edgar Payan (phases SD, DD, CD), Yoshi Uchiyama (construction administration)
Lead Façades Alberto Barba, Maria Rodero (design stages), Kaloyan Erevinov (construction supervision)
Lead Interiors Claudia Glass Dorner (schematic design stages), Mei-Ling Lin (construction supervision)
Lead Landscape Samson Lee
Project Team Arturo Revilla, Muriel Boselli, Chao-Chin Wang, Ai Sato, Michael Harris, Dennis Brezina, Claudia Doner, Mei-Ling Lin, Osbert So, Pierandrea Angius, Diego Perez Espitia, Kaloyan Erevinov, Laurence Dudeney, Albert Ferrer, Leonid Krykhtin, Mu Ren, Kwanphil Cho, Evgeniya Yatsyuk, Lauren Barclay, Henning Hansen, Ben Kikkawa, Nicholette Chan, Michael Grau, Sarah Thurow, Adrian Krezlik, Chiwai Chan, Gordana Jakimovska, Ed Gaskin, Andrea D'imperio, Samson Lee, Shaowen Deng, Will Chen, Joei Kung, Maren Klasking, Carlos Parraga-Botero
Competition Team Ergin Birinci, Maria Tsiorni, Michael Rissbacher, Seda Zirek, Spyridon Kaprinis, Xiaosheng Li
Local Design Institute Siadr – Shanghai Institute of

Architectural Design & Research Ltd
Façade Consultant Thornton Tomasetti
Lighting Lightdesign
Structural Engineering Siadr – Shanghai Institute of Architectural Design & Research Ltd
Mechanical Engineering Parsons Brinckerhoff
Sustainability Consultant Aecom
Gross Floor Area 342,500 m²
Site Area 86,000 m²

Grand Theatre de Rabat [256]
Rabat, Morocco, 2010–
Client Agence pour l'Aménagement de la Vallée du Bouregreg
Design Zaha Hadid with Patrik Schumacher
Project Director Nils-Peter Fischer
Project Team Martin Krcha, Yevgeniya Pozigun, Erwan Gallou, Michail Desyllas, Duarte Reino, Katharina Hieger, Joshua Noad, Thanh Dao
Local Architect Cabinet Omar Alaoui (Morocco)
Structural Engineer Adams Kara Taylor (London)
MEP Engineer Max Fordham (London)
Acoustics & Theatre Artec Consultant (New York)
Façade Newtecnic (London)
Lighting Office for Visual Interaction Inc (New York)
Cost Consultant Donnell Consultants Incorporated (Florida)
Acoustics & Theatre Arup (New York)
Kitchen IR2A (Rabat)
Landscaping PROAP
Local MEP and Structural Consultant Omnium (Rabat)
Fire Consultant Casavigilance (Rabat)
Site Area Approx. 55,000 m²
Gross Floor Area Approx. 25,425 m²

BBK Headquarters [257]
Bilbao, Spain, 2010–
Client BBK (Bilbao Bizkaia Kutxa)
Design Zaha Hadid with Patrik Schumacher
Project Director Manuela Gatto
Technical Associate Dillon Lin
Project Team Teoman N. Ayas, Muriel Boselli, Maren Klasing, Edgar Payan, Arturo Revilla, Ai Sato, Camiel Weijenberg, Seda Zirek
Engineering Consultant Arup Madrid
Gross Floor Area 25,000 m²

One Thousand Museum [257]
Miami, Florida, USA, 2012–
Client 1000 Biscayne Tower, LLC
Design Zaha Hadid with Patrik Schumacher
Project Director Chris Lépine
Project Team Alessio Constantino, Martin Pfleger, Oliver Bray, Theodor Wender, Irena Predalic, Celina Auterio, Carlota Boyer
Competition Team Sam Saffarian, Eva Tiedemann, Brandon Gehrke, Cynthia Du, Grace Chung, Aurora Santan, Olga Yatsyuk
Local Architect O'Donnell Dannwolf Partners
Structural Engineer DeSimone
MEP Engineer HNGS Consulting Engineers
Civil Engineer Terra Civil Engineering
Landscape Enea Garden Design
Fire Protection SLS Consulting Inc
Vertical Transportation Lerch Bates Inc
Wind Tunnel Consultant RWDI
Gross Floor Area 84,637 m²

Tea and Coffee Set [262]
1995–1996
Client Sawaya & Moroni
Design Zaha Hadid
Design Team Maha Kutay, Anne Save de Beaurecueil
Material Stainless steel

Tea and Coffee Piazza [263]
2003
Client Alessi
Design Zaha Hadid with Patrik Schumacher
Design Team Woody Yao, Thomas Vietzke

Z-Scape [264]
2000
Client Sawaya & Moroni
Design Zaha Hadid
Design Team Caroline Voet, Woody Yao, Chris Dopheide, Eddie Can

Iceberg [265]
2003
Client Sawaya & Moroni
Design Zaha Hadid with Patrik Schumacher
Design Team Thomas Vietzke, Woody Yao

Ice Storm [266–267]
2003
Client Österreichisches Museum für Angewandte Kunst
Design Zaha Hadid with Patrik Schumacher
Project Architects Thomas Vietzke, Woody Yao

Belu Bench [268]
2005
Client Kenny Schachter
Design Zaha Hadid with Patrik Schumacher
Project Designer Saffet Kaya Bekiroglu
Project Team Maha Kutay, Tarek Shamma, Melissa Woolford

Zaha Hadid Bowls 60, 70 and Metacrylic [268]
2007
Client Sawaya & Moroni
Design Zaha Hadid with Patrik Schumacher
Lead Designer Saffet Kaya Bekiroglu
Design Team Maha Kutay, Melissa Woolford, Tarek Shamma
Size (Bowl 60) 600 mm (width) × 275 mm (depth) × 130 mm (height)
Size (Bowl 70) 700 mm (width) × 325 mm (depth) × 130 mm (height)
Size (Metacrylic) 700 mm (width) × 325 mm (depth) × 130 mm (height)

Aqua Table [269]
2005
Client Established & Sons
Design Zaha Hadid with Patrik Schumacher
Project Designer Saffet Kaya Bekiroglu
Design Team Tarek Shamma
Size 420 cm × 148.5 cm × 72 cm

Flow [269]
2006–2007
Client Serralunga
Design Zaha Hadid with Patrik Schumacher
Lead Designers Michele Pasca di Magliano, Viviana R. Muscettola

Z-Car I and II [270]
2005–2008
Client Kenny Schachter
Design Zaha Hadid with Patrik Schumacher
Project Designer Jens Borstelmann
Design Team David Seeland
Size 3.68 m (length) × 1.7 m (width) × 1.4 m (height); wheelbase: 2.45 m

Z-Island [271]
2005–2006
Client DuPont Corian
Design Zaha Hadid with Patrik Schumacher
Project Architect Thomas Vietzke
Design Team Georgios Maillis, Maurice Martel, Katharina Neuhaus, Ariane Stracke
Manufacturer Hasenkopf
Size Exhibition space: 214 m²; main island: 4.5 m (length) × 0.8 m (width) × 1.8 m (height); second island: 1.2 m × 1.6 m × 0.9 m; wall panels: 100 pieces at 0.6 m × 0.6 m

The Seamless Collection [272]
2006
Client Established & Sons; Phillips de Pury & Company
Design Zaha Hadid with Patrik Schumacher
Design Team Saffet Kaya Bekiroglu, Melodie Leung, Helen Lee, Alvin Huang, Hannes Schafelner

Dune Formations [273]
2007
Client David Gill Galleries
Design Zaha Hadid with Patrik Schumacher
Design Team Michele Pasca di Magliano, Viviana R. Muscettola
Materials Aluminium, resin
Size 24 m × 13 m

Crater Table [274]
2007
Client David Gill Galleries
Design Zaha Hadid with Patrik Schumacher
Project Designer Saffet Kaya Bekiroglu
Design Team Chikara Inamura, Chrysostomos Tsimourdagkas

Moon System [274]
2007
Client B&B Italia
Design Zaha Hadid with Patrik Schumacher
Design Lead Viviana R. Muscettola
Design Team Michele Pasca di Magliano

Mesa Table [275]
2007
Client Vitra
Design Zaha Hadid with Patrik Schumacher
Project Designer Saffet Kaya Bekiroglu
Project Team Chikara Inamura, Melike Altinisik

Zaha Hadid Chandelier [276]
2008
Client Swarovski Crystal Palace

Design Zaha Hadid with Patrik Schumacher
Project Designers Saffet Kaya Bekiroglu, Kevin McClellan
Project Team Jaime Bartolomé, Simon Koumjian, Amit Gupta
Engineering Consultant Arup AGU: Tristan Simmonds
Fabrication LDDE Vertriebs GmbH; Stainless Steel Solutions; Sheetfabs Ltd
Materials Swarovski crystal, aluminium, SS cabling, copper wire, microprinted LEDs
Size 6 m (height) × 2.25 m (width) × 10.5 m (length)

Swarm Chandelier [276]
2006
Client Established & Sons
Design Zaha Hadid with Patrik Schumacher
Project Architect Saffet Kaya Bekiroglu
Manufacturer Established & Sons
Material Crystal

Vortexx Chandelier [277]
2005
Client Sawaya & Moroni
Design Zaha Hadid with Patrik Schumacher
Design Team Thomas Vietzke
Partners Sawaya & Moroni; Zumtobel Illuminazione s.r.l.
Material Fibreglass, car paint, acrylic, LED
Size 1.8 m (diameter)

Crevasse Vase [278]
2005–2008
Client Alessi Spa
Design Zaha Hadid with Patrik Schumacher
Design Team Woody Yao, Thomas Vietzke
Manufacturer Alessi Spa
Material Silver-plated stainless steel

WMF Cutlery [278]
2007
Client WMF
Design Zaha Hadid with Patrik Schumacher
Project Designer Jens Borstelmann
Material Mirror-polished stainless steel

Series ZH Door Handles [279]
2007
Client Valli & Valli
Design Zaha Hadid with Woody Yao
Material Nikrall Zamak alloy UNI 3717

Icone Bag [280]
2006
Client Louis Vuitton
Design Zaha Hadid with Patrik Schumacher
Project Designer Ana M. Cajiao
Design Team Muthahar Khan

Melissa Shoe [280]
2008
Client Melissa/Grendene S/A
Design Zaha Hadid with Patrik Schumacher
Lead Designer Ana M. Cajiao
Design Team Maria Araya, Muthahar Khan

Lamellae Collection [281]
2016
Client Georg Jensen
Design Zaha Hadid with Patrik Schumacher
Design Team Maha Kutay, Woody Yao, Weilong Xei
Materials Various, including silver, gold, diamonds, rhodium plating
Size Bangle 19 mm (height), 15.2–18.5 cm (circumference, small–large); double ring 28.5 mm; long bangle 9 mm (width), 14.5–17.1 cm (circumference, S/M–M/L); long ring 73 mm × 25 mm; ring I 41 mm × 40 mm; ring II 57 mm × 37.5 mm; twin ring 73 mm × 47.5 mm; twisted bangle 13 cm (length), 17–18.5 cm (S/M–M/L)

Lotus, Venice Biennale [282]
Venice, Italy, 2008
Client La Biennale di Venezia
Design Zaha Hadid with Patrik Schumacher
Design Team Melodie Leung, Gerhild Orthacker
Materials Glass-reinforced plastic and polyurethane with high-gloss lacquer paint finish, foam mattress, wood, printed fabric, synthetic rubber, and stretch fabric
Size Closed: 5.7 m (length) × 5.7 m (width) × 2.6 m (height); Open: 10.2 m (length) × 6.1 m (width) × 2.6 m (height)

Home Bar [283]
London, UK, 2008
Client Home House Ltd
Design Zaha Hadid with Patrik Schumacher
Design Team Maha Kutay, Melissa Woolford, Woody Yao, Susanne Berggen, Sophie Le Bienvenue, Susu Xu, Gabriela Jimenez
Materials Fibreglass, resin, fabric
Total Area 158 m²

Aura [283]
Venice, Italy, 2008
Client Fondazione La Malcontenta
Design Zaha Hadid with Patrik Schumacher
Design Team Fulvio Wirz, Mariagrazia Lanza
Lighting Designer Zumtobel Illuminazione s.r.l.
Manufacturer Idee & Design GmbH; The Art Factory
Core Material PU-Foam RG100 and Polystyrene EPS 20 milled parts; coated with PU-SB1
Installation Details Aura-L and Aura-S: lacquered fibreglass with embedded steel frame
Size 6 m × 3 m × 2.45 m

Cirrus [284]
Cincinnati, Ohio, USA, 2008
Client Lois & Richard Rosenthal Center for Contemporary Art
Design Zaha Hadid with Patrik Schumacher
Design Team Melodie Leung, Gerhild Orthacker
Materials Laminate by Formica®, colour Black (909-90) in polished finish stained medium-density fibreboard
Size 1.8 m (diameter)

Seoul Desk and Table [284]
2008
Client NY Projects
Design Zaha Hadid with Patrik Schumacher
Designer Daniel Widrig
Size 422 cm × 125 cm × 72 cm

Stuart Weitzman Boutiques [285]
Milan, Hong Kong, Rome, 2013–2014
Client Stuart Weitzman
Design Zaha Hadid with Patrik Schumacher
Milan:
Project Architect Paola Cattarin
Design Team Alessio Costantino, Maren Klasing, Jorge Mendes-Caceres, Vincenzo Barilari, Zetta Kotsioni
Local Executive Architects Milano Layout, Marco Claudi, Sara Acconcia
Structural Engineer Bruni Salesi
Lighting Pollice Illuminazione
Contractor Tecnolegno Allestimenti, Claudio Radice
Hong Kong:
Project Architect Paola Cattarin
Design Team Alessio Costantino, Maren Klasing, Jorge Mendes-Caceres, Vincenzo Barilari, Zetta Kotsioni, Hinki Kuong
Local Executive Architects CW Tang
Client Project Manager Wong Chi Lap
Lighting Twinsen Ho
Contractor Vitus Man
Rome:
Project Architect Paola Cattarin
Design Team Jorge Mendes-Caceres, Vincenzo Barilari, Zetta Kotsioni, Soungmin Yu, Kyle Dunnington
Local Executive Architects Milano Layout, Marco Claudi, Sara Acconcia
Structural Engineer Ing. Bruno Zeuli
Lighting Pollice Illuminazione
Craftsmen Me Cubo, Marco Marelli
Contractor Tecnolegno Allestimenti s.r.l., Claudio Radice

Table, Shelf and Henry Moore Exhibition Design [286]
2008
Client Hauser & Wirth
Design Zaha Hadid with Patrik Schumacher
Project Team Woody Yao, Dylan Baker-Rice
Material Aluminium
Finish Mirror Finish
Dimensions:
Length 6.42 m
Width 2.18 m
Height 1.04 m

Scoop Sofa [286]
2008
Client Sawaya & Moroni and ROVE LLP
Design Zaha Hadid with Patrik Schumacher
Design Team Saffet Kaya Bekiroglu, Melodie Leung, Maha Kutay, Dylan Baker-Rice, Filipa Gomes
Materials Glass reinforced plastic (GRP), with pearlized lacquered paint finish
Dimensions:
Length 3.90 m
Depth 1.41 m
Height 0.83 m

Kloris [287]
2008
Outdoor seating
Client Julian Treger and Kenny Schachter/Rover Gallery
Design Zaha Hadid with Patrik Schumacher
Design Team Melodie Leung, Tom Wuenschmann, Yael Brosilovski
Materials Glass reinforced plastic (GRP), with high-gloss lacquer finish in chrome and gradations of green, steel

base plates
Dimensions 6500 mm × 5100 mm × 80 mm

Glace Collection [287]
2009
Client Swarovski
Design Zaha Hadid with Patrik Schumacher
Design Team Swati Sharma, Maria Araya
Dimensions:
Necklace 220 mm × 140 mm
Cuff 1 105 mm × 50 mm
Cuff 2 110 mm × 50 mm
Ring 1 45 mm × 49 mm
Ring 2 70 mm × 35 mm
Pendant 150 mm × 110 mm
Materials Coloured resin and Swarovski crystals in jet, opal, crystal, padparadscha and black diamond

J. S. Bach Music Hall [288–289]
Manchester, UK, 2009
Client Manchester International Festival
Design Zaha Hadid with Patrik Schumacher
Design Team Melodie Leung, Gerhild Orthacker
Acoustic Consultant Sandy Brown Associates
Tensile Structural Engineer Tony Hogg Design Ltd
Fabricator Base Structures
Fabric Trapeze Plus Lycra (approximately 650 m²)
Lighting DBN Lighting Limited
Size 17 m × 25 m

Avilion Triflow Taps [290]
2009
Client Avilion Triflow
Design Zaha Hadid with Patrik Schumacher
Project Architects Woody Yao, Dylan Baker-Rice, Maha Kutay, Melissa Woolford
Materials Chrome-plated brass-finish chrome, mirror-image finish
Dimensions:
Kitchen Length 43 cm, width 27 cm, height 34 cm
Bath Length 40 cm, width 23 cm, height 20 cm

Skein Sleeve Bracelet [290]
2009
Client Sayegh Jeweller
Design Zaha Hadid with Patrik Schumacher
Design Team Maha Kutay, Melissa Woolford, Michael Grau, Hussam Chakouf

Genesy Lamp [291]
2009
Client Artemide
Design Zaha Hadid with Patrik Schumacher
Project Architect Alessio Costantino
Material Expanded Polyurethane lacquered
Finish Gloss
Dimensions:
Height 195 cm
Length 120 cm
Width 59 cm

Tide [291]
2010
Client Magis
Design Zaha Hadid with Patrik Schumacher

Design Team Ludovico Lombardi, Viviana Muscettola, Michele Pasca di Magliano
Finish Gloss
Dimensions:
Free-standing Module S Width 450 mm; depth 450 mm; height 450 mm
Free-standing Module L Width 450 mm; depth 450 mm; height 1,350 mm
Wall Module S Width 450 mm; depth 275 mm; height 450 mm
Wall Module L Width 1,350 mm; depth 275 mm; height 450 mm
Material Liquid wood

Zaha Hadid: Architects & Suprematism [292]
Zurich, Switzerland, 2010
Client Galerie Gmurzynska
Design Zaha Hadid with Patrik Schumacher
Project Director Woody Yao
Project Architect Melodie Leung
Project Team Maha Kutay, Manon Janssens, Filipa Gomes, Aram Gimbot
Artist Consultant Antonio De Campos
Gallery Directors Krystyna Gmurzynska, Mathias Rastorfer
Area 230 m²

Zaha Hadid: Palazzo della Ragione Padova [293]
Padua, Italy, 2009–2010
Client Fondazione Barbara Cappochin
Design Zaha Hadid with Patrik Schumacher
Exhibition Team Viviana Muscettola, Michele Pasca di Magliano, Woody Yao, Elif Erdine
Installation Coordination Manon Janssens, Woody Yao, Filipa Gomez, Shirley Hottier
Main Contractor Idee & Design
Lighting Design I-Guzzini

Roca London Gallery [293]
London, UK, 2009–2011
Client Roca
Design Zaha Hadid with Patrik Schumacher
Design Team Melodie Leung, Gerhild Orthacker
Project Directors Woody Yao, Maha Kutay
Project Architect Margarita Yordanova Valova
Design Development Gerhild Orthacker, Hannes Schafelner, Jimena Araiza, Mireia Sala Font, Erhan Patat, Yuxi Fu, Michal Treder, Torsten Broeder
Concept Design Dylan Baker-Rice, Melissa Woolford, Matthew Donkersley, Maria Araya
Structural and Façade Engineering Buro Happold
MEP and Acoustics Consultant Max Fordham Consulting Engineers
Lighting Design Isometrix Lighting + Design
AV Consultant Sono
Cost Manager Betlinski
Construction Manager Empty, S.L.
Area 1,100 m²

Z-Chair [294]
2011
Client Sawaya & Moroni
Design Zaha Hadid with Patrik Schumacher
Design Team Fulvio Wirz, Mariagrazia Lanza, Maha Kutay, Woody Yao
Dimensions:

Length 920 mm
Height 880 mm
Depth 610 mm
Materials Polished stainless steel

Art Borders by Zaha Hadid [295]
2010
Client Marburg Wallcoverings
Design Zaha Hadid with Patrik Schumacher
Project Leader Melodie Leung
Project Team Filipa Gomes, Danilo Arsic, Russel Palmer, Salvatore Lillini, Maha Kutay, Woody Yao
Marburg Art Director Dieter Langer

Zaha Hadid: Fluidity & Design [296]
Muharraq, Bahrain, 2010
Client Shaikh Ebrahim bin Mohammed Al Khalifa Centre for Culture and Research
Design Zaha Hadid with Patrik Schumacher
Project Team Elke Frotscher, Woody Yao, Filipa Gomes, Manon Janssens, Melodie Leung, Maha Kutay
Artist Consultant Antonio De Campos

Zaha Hadid: Une Architecture [296]
Paris, France, 2011
Client Institut du Monde Arabe
Design Zaha Hadid with Patrik Schumacher
Exhibition Team Thomas Vietzke, Manon Janssens, Woody Yao, Jens Borstelmann, Sofia Daniilidou, Torsten Broeder, Tiago Correia, Danilo Arsic, Victor Orive, Mostafa El Sayed, Christoph Wunderlich, Martin Krcha, Niki Berry, Claudia Fruianu, Daniel Widrig, Filipa Gomes
Artist Consultant Antonio De Campos
Production AIA Production (APC+AIA): Renaud Sabari, Alexandra Cohen
Visual Identity Yorgo Tloup
Video Projections Cadmos
Production Coordination Emmanuel Lemercier de L'Ecluse
Pavilion Installation:
Artistic Supervision Zaha Hadid Architects: Thomas Vietzke, Jens Borstelmann
Client's Assistant AIA Productions
Project Management Elioth Iosis and ENIA
Building Viry (Fayat Group); Satelec (Fayat Group); Extenzo
Gross Floor Area 700 m² (29 m × 45 m)
Exhibition space 500 m²

Zaha Hadid: Form in Motion [297]
Philadelphia, Pennsylvania, USA, 2011–2012
Client Philadelphia Museum of Art
Design Zaha Hadid with Patrik Schumacher
Project Director Woody Yao
Exhibition Design Team Jimena Araiza, Filipa Gomes
Exhibition Coordinators Manon Janssens, Maha Kutay
Area 395 m²

Zaha Hadid: Parametric Tower Research [297]
Cologne, Germany, 2012
Client AIT ArchitekturSalons
Design Zaha Hadid with Patrik Schumacher
Exhibition Team Thomas Vietzke, Manon Janssens, Woody Yao, Jens Borstelmann, Sofia Daniilidou, Torsten Broeder, Anna Roeder
Artist Consultant Antonio De Campos
Area 300 m²

Twirl [298]
2011
Client Lea Ceramiche and Interni
Design Zaha Hadid with Patrik Schumacher
Project Director Woody Yao
Project Architect Johannes Schafelner
Team Yuxi Fu, Manon Janssens, Maha Kutay, Danilo Arsic, Filipa Gomes
Area 800 m^2

Floating Staircase [298]
London, UK, 2012
Design Zaha Hadid with Patrik Schumacher
Design Lead Melodie Leung
Design Team Garin O'Aivazian, Bear Shen
Structural Consultant Adams Kara Taylor: Christian Tygoer
Fabrication Il Cantiere
Structural Engineer C&E Ingeniere: Raphaël Fabbri
Steelwork and Installation Artistic Engineering
Size 8.8 m (length) × 2.5 m (width) × 3.2 m (height)

Arum Installation and Exhibition [299]
Venice, Italy, 2012
Client Venice Architecture Biennale
Design Zaha Hadid with Patrik Schumacher
Exhibition Design Woody Yao, Margarita Valova
Installation Design and Presentation Shajay Bhooshan, Saman Saffarian, Suryansh Chandra, Mostafa El Sayed
Structural Engineering Buro Happold: Rasti Bartek
Material & Fabrication Technology RoboFOLD: Gregory Epps
Coordinator Manon Janssens
Collaborators Studio Hadid, Universität für angewandte Kunst: Jens Mehlan, Robert Neumayr, Johann Traupmann, Christian Kronaus, Mascha Veech, Mario Gasser, Susanne John; THE BLOCK Research Group, Institute of Technology in Architecture, ETH Zurich: Philippe Block, Matthias Rippmann; Faculty of Architecture, ETH Zurich: Toni Kotnik; Centro de Investigaciones y Estudios de Posgrado, Faculty of Architecture, UNAM, Mexico: Juan Ignacio del Cueto Ruiz-Funes
With the support of Permasteelisa Spa ARTE & Partners
Size 2.8 m × 2.8 m (base); 10 m × 7.8 m (top); 5.8 m (height)

Citco Tau Vases and Quad Tables [300]
Milan, Italy, 2015
Client Citco Italia
Design Zaha Hadid with Patrik Schumacher
Design Team Woody Yao, Maha Kutay, Sara Saleh, Filipa Gomes, Niran Buyukkoz
Material Marble
Tau Vases Bardiglio Nuvolato (extra small & extra large), Statuario (small), Bianco Covelano (medium), Bianco Carrara (large)
Quad Tables Statuario (small 1), Nero Marquina (small 2), Silver Wave (medium), Nero Assoluto (extra large & console)
Dimensions:
Tau Vases extra small (34 cm × 34 cm × 20 cm), small (43 cm × 43 cm × 20 cm), medium (40 cm × 40 cm × 34 cm), large (40 cm × 40 cm × 48 cm), extra large (50 cm × 50 cm × 74 cm)
Quad Tables small 1 (60 cm × 60 cm × 45 cm), small 2 (60 cm × 60 cm × 45 cm), medium (90 cm × 90 cm × 35 cm), extra large (150 cm × 150 cm × 27 cm), console (160 cm × 45 cm × 90 cm)

Liquid Glacial Dining & Coffee Table [301]
2012
Client David Gill Galleries
Design Zaha Hadid with Patrik Schumacher
Design Team Fulvio Wirz, Mariagrazia Lanza, Maha Kutay, Woody Yao
Project Director Woody Yao
Project Architect Maha Kutay
Dining Table:
Dimensions Section 1 Length 2,515 mm, Depth 1,402 mm, Height 750 mm; Section 2 Length 2,827 mm, Depth 1,405 mm, Height 750 mm
Material Polished Plexiglas Clear
Limited Edition 8 (+ 2 Artist Proofs + 2 Prototypes)
Coffee Table:
Dimensions Section 1 Length 2,500 mm, Depth 870 mm, Height 400 mm
Material Polished Plexiglas Clear
Limited Edition 8 (+2 Artist Proofs + 2 Prototypes)

Zaha Hadid Retrospective [302–303]
'Zaha Hadid at the State Hermitage Museum', Nicolaevsky Hall (Winter Palace), State Hermitage Museum, St Petersburg, Russia, 27 June – 27 September 2015
Exhibition Design Zaha Hadid with Patrik Schumacher
Project Directors Woody Yao, Maha Kutay
Exhibition Coordinator Manon Janssens
Design Team Johanna Huang, Daria Zolotareva, Filipa Gomes, Olga Yatsyuk, Margarita Valova, Zahra Yassine, Jessika Green
Video Installation Henry Virgin
The State Hermitage Museum Dr Mikhail Piotrovsky (General Director), Ksenia Malich (Exhibition Curator), Dr Dimitri Ozerkov (Director of the Contemporary Art Department), Geraldine Norman (Advisor to the Director), Vitaly Korolev, Boris Kuziakin

---

Photographs of models, paintings and line drawings by Edward Woodman unless otherwise stated.

Hufton + Crow Photographers 159 (right, top and bottom), 171 (top right and centre left), 180–181, 182, 183, 184–185, 210 (top left, top right and bottom right), 224, 225, 240 (top left, bottom left and bottom right), 252–253, 254–255, 293 (bottom); Iwan Baan 4–5, 107–109, 156–157, 160 (bottom), 171 (top left), 174 (bottom), 175 (right, top and bottom), 188 (top right), 220 (top left, top right and bottom left), 242 (top), 243 (bottom left and right), 244–245, 250–251, 293 (bottom); Luke Hayes 16, 178 (top), 179 (top left), 190 (bottom), 191, 192 (all except top right), 193, 198–199, 268 (top), 276 (bottom), 282, 283, 286 (top), 298 (top left); Richard Bryant/Arcaid 29 (bottom right), 118 (left, top and bottom); Christian Richters 33 (bottom), 52–55, 94–95, 158–159, 188 (bottom right), 189; © Paul Warchol 46–47, 63, 226–227, 297 (top); Hélène Binet 48, 78, 83, 88–91, 99 (bottom right), 100, 104–105, 114–115, 122, 123, 126, 127 (top and bottom), 128, 129, 132–133, 135 (right, top and bottom), 139 (left and bottom right), 141 (bottom left), 142–143, 144 (top), 146–148, 152 (bottom left), 154 (top), 166–167, 216 (bottom), 217 (bottom, left and right), 239, 297 (top); Herman Van Doorn 56–57; Bruno Komflar 74–75; Margherita Spiluttini 76–77; Markus Doschanti 84 (top); Roland Halbe 98, 99 (top and bottom left), 103 (bottom), 134–135, 136–137, 168–169 (centre), 246; Richard Rothan 103 (top); Werner Huthmacher 116–117, 118–119 (top and bottom), 120–121, 140, 141 (top left), 168 (left, top and bottom), 280 (top); Fernando Guerra 139 (top right), 176–177, 178 (bottom, left and right), 179 (top right and bottom); courtesy of Alessi 145, 263, 278 (top); courtesy of Zaha Hadid Architects 152 (top), 284 (top), 287, 288 (bottom), 291 (top), 298 (bottom); courtesy of Silken Hotels 155; Virgile Simon Bertrand 160 (top), 161, 214 (bottom, left and right), 215 (top left and bottom); © Virgile Simon Bertrand/Alphaville. hk 200, 201, 202–203, 204, 205, 247 (top left and top right); McAteer Photograph/Alan McAteer 170, 171 (centre right); David Bombelli 174 (top); courtesy of City Life 175 (left); Mathias Schormann 190 (top); Frener + Reifer 192 (top right); renders by VA © Zaha Hadid Architects 207; photo © inexhibit.com 208 (top), 209 (bottom left); photo © Wisthaler.com 209 (top); Aaron Pocock 210 (centre right); Toshio Kaneko 212–213, 214 (top left), 215 (centre); Marc Gerritson 214 (top right); Ferid Xayruli 216 (top), 217 (top); doublespace photography 221, 222–223; renders by MIR 228 (centre right and bottom right), 229, 234 (top right and centre right), 235 (top and bottom right), 236–237; renders by VisualArch 228 (top left and top right), 230–231; Tim Fisher 238 (top left), 240 (top right and centre right), 241 (right); photograph by Julian Faulhaber © DACS 2017 249 (top); courtesy of Sawaya & Moroni 262, 264, 265, 268 (bottom), 269 (bottom), 286 (bottom); Gerald Zugmann 266–267; David Sykes 269 (top); courtesy of Kenny Schachter/ROVE 270 (left, top and centre); courtesy of DuPont Corian 271 (top); Leo Torri, courtesy of DuPont Corian 271 (bottom); Jack Coble 272; courtesy of ORCH 273, 293 (top); Michael Molloy 274 (top); Fabrizio Bergamo 274 (bottom); Babara Sorg 275 (top); Leo Torri 276 (top); courtesy of Zumtobel 277; courtesy of WMF 278 (bottom); courtesy of Valli & Valli 279; courtesy of Melissa 280 (bottom); courtesy of Georg Jensen 281 (top left, top right and centre); © Christian Hogstedt 281 (bottom left); Jacopo Spilimbergo 285 (left and top right), 301; Ruud Van Gessel 288 (top), 289; courtesy of Tony Hogg Design and Base Structures 288 (centre); courtesy of Triflow Concepts 290 (top); Will Thom 290 (bottom); courtesy of Magis 291 (bottom); Martin Ruestchi 292; Ruy Teixera, courtesy of Sawaya & Moroni 294; Konstantin Eulenburg 295; Picture Arabia, courtesy of the Shaikh Ebrahim Center, Bahrain 296 (top); François Lacour, courtesy of Institut du Monde Arabe 296 (bottom); Jochen Stueber 297 (bottom); courtesy of Lea Ceramiche @ Andrea Martiradonna 298 (top left); Sergio Pirrone 299 (bottom); Zaha Hadid at the State Hermitage Museum © Yuri Molodkovets 302 (top and bottom), 303

索引

24 Cathcart Road 29
42nd Street Hotel 79
59 Eaton Place 21
582–606 Collins Street 207
Abu Dhabi Performing Arts Centre 232–233
Al Wahda Sports Centre 37
Albertina Extension 130
A New Barcelona 42
Aqua Table 269
Art Borders by Zaha Hadid 295
Arthotel Billie Strauss 65
Arum Installation and Exhibition 299
Aura 283
Avilion Triflow Taps 290
Azabu-Jyuban 34
BBC Music Centre 152
BBK Headquarters 257
Bee'ah Headquarters 234–237
Belu Bench 268
Bergisel Ski Jump 114–115
Berlin 2000 39
Blueprint Pavilion, Interbuild 95 78
BMW Plant Central Building 132–135
Boilerhouse Extension 82
British Pavilion, Venice Biennale 123
Burnham Pavilion 246
California Residence 164
Campus Center, Illinois Institute of Technology 96
Capital Hill Residence 186–187
Car Park and Terminus Hoenheim-Nord 102–105
Cardiff Bay Opera House 70–73
Carnuntum 68–69
Centro JVC Hotel 125
Cirrus 284
Citco Tau Vases & Quad Tables 300
City Life Milano 172–175
City of Towers, Venice Biennale 145
CMA CGM Headquarters Tower 188–189
Concert Hall, Copenhagen 65
Crater Table 274
Crevasse Vase 278
Danjiang Bridge 228–231
Desire 152
Deutsche Guggenheim 190
D'Leedon 210
Dominion Office Building 224–225
Dongdaemun Design Plaza 200–205
Dubai Opera House 196–197
Dune Formations 273
Dutch Parliament Extension 19
Eleftheria Square Redesign 165
Eli & Edythe Broad Art Museum 226–227
Evelyn Grace Academy 198–199
Floating Staircase 298
Flow 269
Galaxy Soho 250–251
Genesy Lamp 291
Glace Collection 287
Grand Buildings, Trafalgar Square 25–27
Grand Theatre de Rabat 256
Great Utopia, The 63
Guangzhou Opera House, 156–161
Guggenheim Museum Taichung 153
Habitable Bridge 84–85
Hafenstraße Development 44–45
Hague Villas, The 62
Halkin Place 28
Hamburg Docklands 30
Heydar Aliyev Centre 216–217
High-Speed Train Station Napoli-Afragola 162–163
Home Bar 283
Hommage à Verner Panton 261
Hotel and Residential Complex 58

Hotel Puerta America – Hoteles Silken 155
IBA-Block 2 33
Ice Storm 266–267
Iceberg 265
Icone Bag 280
Interzum 91 59
Investcorp Building 192–193
Irish Prime Minister's Residence 20
Jockey Club Innovation Tower 220–223
J. S. Bach Music Hall 288–289
Kartal-Pendik Masterplan 194–195
King Abdullah Petroleum Studies & Research Centre 248–249
King Abdullah Financial District Metro Station 211
Kloris 287
Kunsthaus Graz 124
Kurfürstendamm 70 32
Kyoto Installations 28
La Fenice 86
La Grande Mosquée de Strasbourg 125
Lamellae Collection, Georg Jensen 281
Landesgartenschau 1999 88–91
Leicester Square 49
Library & Learning Centre, Vienna 242–245
Lilas 191
Liquid Glacial Dining & Coffee Table 301
Lois & Richard Rosenthal Center for Contemporary Art 97–99
London 2066 60–61
London Aquatics Centre 180–185
López de Heredia Pavilion 142–143
Lotus, Venice Biennale 282
Lycée Français Charles de Gaulle 81
Maggie's Centre Fife 140–141
Malevich's Tektonik 18
Master's Section, Venice Biennale 84
MAXXI: National Museum of XXI Century Arts 106–109
Melbury Court 24
Melissa Shoe 280
Mesa Table 275
Meshworks 122
Messner Mountain Museum Corones 208–209
Metapolis, Charleroi/Danses 112
Metropolis 38
Mind Zone, The 100
Mobile Art Pavilion for Chanel 212–215
Monographic Exhibition 144
Monsoon Restaurant 46–47
Moon System 274
Museum Brandhorst 149
Museum for the Royal Collection 110
Museum of Islamic Arts 92–93
Museum of the Nineteenth Century 19
Music-Video Pavilion 56–57
National Library of Quebec 113
New York, Manhattan: A New Calligraphy of Plan 31
Nordpark Cable Railway 166–169
NYC 2012 Olympic Village 154
One-North Masterplan 131
One Thousand Museum 257
'Opus' Office Tower 218–219
Ordrupgaard Museum Extension 136–139
Osaka Folly, Expo1990 48
Pancras Lane 81
Parc de la Villette 21
Peak, The 22–23
Pet Shop Boys World Tour 101
Phæno Science Centre 116–121
Philharmonic Hall, Luxembourg 87
Pierres Vives 146–148
Port House 238–241
Price Tower Arts Center 144
Red Sofa 260
Regium Waterfront 206
Reina Sofía Museum Extension 111

Rheinauhafen Redevelopment 66–67
Riverside Museum 170–171
Roca London Gallery 293
Rothschild Bank Headquarters and Furniture 111
Royal Palace Hotel and Casino 112
Salerno Maritime Terminal 126–129
Scoop Sofa 286
Seamless Collection, The 272
Seoul Desk and Table 284
Series ZH Door Handles 279
Serpentine Gallery Pavilion 123
Sheikh Zayed Bridge 94–95
Skein Sleeve Bracelet 290
Sky Soho 252–255
Snow Show, The 154
Spittalmarkt 80
Spittelau Viaducts 74–77
Stuart Weitzman Boutiques 285
Swarm Chandelier 276
Table, Shelf and Exhibition Design 286
Tea and Coffee Piazza, Alessi 263
Tea and Coffee Set, Sawaya & Moroni 262
Temporary Museum, Guggenheim Tokyo 131
Tents and Curtains, Milan Triennale 29
Tide 291
Tokyo Forum 43
Tomigaya 35
Twirl 298
UNL/Holloway Road Bridge 101
Urban Nebula 190
Victoria City Areal 40–41
Vision for Madrid 64
Vitra Fire Station 52–55
Vortexx Chandelier 277
Vorwerk Wall-To-Wall Carpeting 261
Waecthenberg Ceramics 260
Wangjing Soho 247
Warped Plane Lamp 260
Wave Sofa 260
West Hollywood Civic Center 36
Whoosh Sofa 260
Wish Machine: World Invention 83
WMF Cutlery 278
World (89 Degrees), The 24
Zaha Hadid: Architects & Suprematism 292
Zaha Hadid Bowls 60, 70 and Metacrylic 268
Zaha Hadid Chandelier 276
Zaha Hadid: Fluidity & Design 296
Zaha Hadid: Form in Motion 297
Zaha Hadid Lounge 130
Zaha Hadid: Palazzo della Ragione Padova 293
Zaha Hadid: Parametric Tower Research 297
Zaha Hadid Retrospective 302–303
Zaha Hadid: Une Architecture 296
Zaragoza Bridge Pavilion 176–179
Z-Car I and II 270
Z-Chair 294
Z-Island 271
Zollhof 3 Media Park 50–51
Zorrozaurre Masterplan 150–151
Z-Scape 264